THE PATH OF N

THE STORY OF THE REVOLUTIC

SIMON AND SCHUSTE

BRUCE SCHECHTER

ESISTANCE

SUPERCONDUCTIVITY

YORK   LONDON   TORONTO   SYDNEY   TOKYO

Simon and Schuster
Simon & Schuster Building
Rockefeller Center
1230 Avenue of the Americas
New York, New York 10020

Designed by Karolina Harris
Manufactured in the United States of America
10 9 8 7 6 5 4 3 2 1

Library of Congress Cataloging in Publication Data

Schechter, Bruce.
 The path of no resistance: the story of the revolution in
superconductivity/Bruce Schechter.
    p. cm.
  Bibliography: p.
  Includes index.
    1. Superconductivity. I. Title.
QC611.95.S34 1989
537.6′23—dc19   89-4190
                                                                CIP

ISBN 0-671-65785-2

The author is grateful for permission to reprint and excerpt from:
The Collected Poems of Wallace Stevens, copyright © 1942 by Wallace
Stevens and renewed 1970 by Holly Stevens, reprinted with the
permission of Alfred A. Knopf, New York

ACKNOWLEDGMENTS

I would like to join thousands of scientists around the world in expressing my gratitude to Alex Müller and Georg Bednorz. Because of their tenacity and courage our lives have truly changed.

In a time when it seemed breakthroughs occurred every day, many scientists nevertheless generously granted me interviews. Among them were Philip Anderson, James Ashburn, Phillipe Barboux, Georg Bednorz, Robert Beyers, Kent Bowen, Marvin Cohen, William Gallagher, Theodore Geballe, Paul Grant, Laura Greene, Richard Greene, Mike Hahn, Alan Hermann, Chao Huang, Koichi Kitazawa, Brian Maple, Alex Müller, Kumar Patel, Rustum Roy, Shoji Tanaka, Jean-Marie Tarascon, James Wong, and Alex Zettl. It was also a time when there seemed to be a new conference every week, and I would like to thank the organizers and the many speakers who generously granted me access to these meetings.

Thanks also to Anne Kottner and Caroline Chapman of the American Institute of Physics; Gerald Present and Christine Anselmi of IBM; Mike Jacobs of AT&T Bell Laboratories; Hiroshi Ishikawa of the Japanese Foreign Press

Center; Kevin Ott of the Council on Superconductivity for American Competitiveness; Paul Maxwell of the Science, Space and Technology Committee of the U.S. House of Representatives; and Gloria Lubkin and Irwin Goodwin of *Physics Today* for numerous kindnesses. John Langone was remarkably generous. Janet Byrne and Stella Dong helped with some of the research.

I was particularly lucky to have the best agent in the world, Kristine Dahl, who introduced me to Bob Bender, the best editor. I was also lucky in my friends, David Cooper, Molly Rauber, and Susannah Greenberg, and luckiest to have had the advice, encouragement, and every sort of aid from my brother, Robert Schechter.

TO THE MEMORY OF MY MOTHER,

SYLVIA SCHECHTER

CONTENTS

José Arcadio Buendía, without understanding, stretched out his hand toward the cake, but the giant moved it away. "Five reales more to touch it," he said. José Arcadio Buendía paid them and put his hand on the ice and held it there for several minutes as his heart filled with fear and jubilation at the contact with mystery. Without knowing what to say, he paid ten reales more so that his sons could have that prodigious experience. Little José Arcadio refused to touch it. Aureliano, on the other hand, took a step forward and put his hand on it, withdrawing immediately. "It's boiling," he exclaimed, startled. But his father paid no attention to him. Intoxicated by the evidence of the miracle, he forgot at that moment about the frustrations of his delirious undertakings . . . He paid another five reales and with his hand on the cake, as if giving testimony on the holy scriptures, he exclaimed: "This is the great invention of our time."

—*One Hundred Years of Solitude*, Gabriel García Márquez

CHAPTER 1

SURMISES, JEALOUSIES, CONJECTURES

Rumor is a pipe
Blown by surmises, jealousies, conjectures . . .

—Shakespeare

As would happen so often in the year ahead, rumor had outraced Koichi Kitazawa. It wasn't a fair race; Kitazawa was only traveling by jet.

He left Tokyo on Friday, the 28th of November 1986, bound for Boston, where he was scheduled to give a talk at the fall meeting of the Materials Research Society. The subject of the talk, which had been fixed for months, was research he and his colleagues had been carrying out for the past eight years—good, solid, interesting work that had earned Kitazawa and his group at Tokyo University an international reputation, and which interested him now barely at all. In the past month the Japanese scientists had unceremoniously, and without a backward glance, abandoned that line of research, abandoned their families and friends, regular meals and sleep, to devote themselves to the study of a new material, dark and brittle as Limoges china.

The fragile new material, which had been discovered earlier that year by two researchers in Switzerland, showed evidence of a nearly mystical property known to physicists as superconductivity. Ever since it was discovered more than seventy-five years ago, superconductivity has seemed almost too good to be true. Unlike ordinary conductors, which exact a stiff energy toll on the electric current they carry, superconductors carry electricity absolutely free. It's the closest thing to perpetual motion the universe has to offer, and although it took physicists until 1957 before they understood how such a thing was possible, without even thinking very hard they came up with dozens of practical applications of superconductors: Superconducting power lines, for example, could save the Niagaras of electrical power that are now thrown away in the form of waste heat. Levitating trains floating on the powerful fields of superconducting magnets could travel between cities at upward of 300 miles per hour. Superconductors could also make computers faster and smaller, they could greatly improve medical scanners, and so on. Wherever energy was wasted—and that means everywhere—superconductors could work miracles. Of course, there was a catch.

Before they shed their electrical resistance all known superconductors must be cooled to nearly absolute zero, as cold as matter can ever get, zero degrees on the Kelvin scale, written 0 K (and equivalent to minus 273 degrees Celsius, or minus 460 degrees Fahrenheit). This is colder than outer space; it is the temperature at which even atoms cease their restless motion. To achieve the glacial temperatures required by known superconductors was both difficult and expensive, requiring liquid helium, which is extremely costly and cumbersome to use. Maintaining these low temperatures requires elaborate, expensive, and bulky insulation. So the wonderful applications of superconductivity that scientists and engineers had dreamed of were mostly too expensive to be practical. The scientists felt like starving people who must feast on beautiful photographs of food.

The obvious thing to do was to try to find a superconductor that worked at a higher temperature. A superconductor that operates at room temperature would be ideal, but even one that required less heroic refrigeration would be an important discovery. After an Edisonian odyssey of trial and error, testing tens of thousands of chemical compounds, the highest-temperature superconductor that anyone had found still had to be cooled to about 23 degrees Kelvin, that is, 23 degrees Celsius above absolute zero. This material, a mixture of the elements niobium and germanium, was discovered in 1971, and had not been improved upon since. Occasionally a report would surface in the literature that the 23-K barrier had been surpassed. But these results always lacked the scientific *sine qua non* of reproducibility. Many physicists had even begun to suspect that 23 K represented a natural limit, like the speed of light, somehow built into the inflexible laws of the universe; some even believed they could prove this mathematically. The message that Kitazawa carried to Boston was that his laboratory had verified, beyond a doubt, that the black ceramic discovered by two Swiss researchers became superconducting at a temperature somewhere around 30 K, that is, 30 degrees Celsius above absolute zero, or approximately minus 400 degrees Fahrenheit, almost torrid by the standards of Kitazawa's profession.

When Kitazawa checked into Boston's Copley Hotel, he was unaware that overnight he had become something of a scientific celebrity. On the day he left Tokyo the prominent Japanese daily *Asahi Shimbun* ran an article about the work Kitazawa, his boss, Shoji Tanaka, and their colleagues at the University of Tokyo were doing. The details were scant—the Swiss researchers who had made the original discovery were not named—but the import of the discovery was not missed by those who read it. Within hours of its publication in Tokyo the international edition was being scanned by Japanese readers all around the world.

That weekend in Zürich, Emi Takashige saw the story and at once phoned her husband, Masaaki Takashige, a research scientist at the IBM Zürich Research Laboratory in the suburb of Rüschlikon. Isn't this what Alex and Georg have been working on, she asked. Alex Müller and Georg Bednorz were the anonymous Swiss scientists mentioned in the story, but they were not to be anonymous for long. By Monday, December 1, the Japanese article had been read, translated, telephoned, and telefaxed by the far-flung Japanese scientific community. Kitazawa could not walk ten steps at the Boston MRS meeting without somebody's stopping him to ask, Is it true? Kitazawa, who had been expelled from graduate school as a radical student leader, enjoyed imparting his revolutionary news. Yes, it was true, he said, and he had the graphs to prove it.

One of the first to stop Kitazawa was Ted Geballe, a physical chemist from Stanford University, who is one of the fathers of modern superconductivity research. He is a tall, thin man in his mid-sixties, with a shock of white hair, bushy, dark eyebrows, and sharp green eyes. Together with his colleague, the late Bernd Matthias, Geballe had discovered many of the superconductors known today. He had also thrown cold water on many of the claims of high-temperature (above 23 K) superconductivity that regularly appeared in the scientific literature. Long ago Geballe had learned scientific caution, a form of enlightened cynicism: When a result seems almost too good to be true, it almost always is.

Before coming to Boston Geballe had also noticed Müller and Bednorz's paper, which had appeared in *Zeitschrift für Physik (Journal of Physics)*, a small scientific journal that has rarely been the forum for announcing major discoveries. And the title of the paper was not designed to arouse enthusiasm either: "Possible High T_c Superconductivity in the Ba-La-Cu-O System." T_c (pronounced "tee-cee") is scientific shorthand for critical temperature, the temperature at which a material becomes superconducting; high T_c was the grail of superconductivity. But all the authors were claiming was "possible" high T_c, which would not be enough to cause many scientists to drop what they were doing and mix up a batch themselves. The paper was equally cautious. Moreover, to Geballe, who knew almost all the scientists working in the field of superconductivity—there weren't that many—the authors' names were unfamiliar. So, instead of heading for the supply room to get the chemicals necessary to make Bednorz and Müller's compound (oxides of barium, lanthanum, and copper), Geballe reached for one of the most valuable instruments of modern science, the telephone, and began making inquiries.

Geballe called friends around the United States, and one in Switzerland. None knew any more than what was written in Bednorz and Müller's paper. He called a friend at AT&T Bell Laboratories, in Murray Hill, New Jersey. Together Geballe and his friend decided that the best way to confirm the Swiss results was to grow crystals of the new material, but that would be very difficult. They decided to wait and see what they could learn at the MRS meeting in Boston.

After talking to Kitazawa in the lobby of the Copley Hotel, Geballe headed directly for the nearest telephone to call his lab and tell them to start ordering the chemicals and whatever else was necessary to make a batch of Bednorz and Müller's new material.

On Wednesday, December 3, Paul Chu, an ambitious young experimental physicist from the University of Houston, arrived at the meeting to give a talk. He is a small, energetic man with a mop of black hair and large glasses that make his head look too large for his body. He was scheduled to present a paper that day, following Kitazawa, and he planned to leave that night. It was his style to breeze into a scientific meeting, give a talk, and breeze out, perhaps on the theory that a moving target is hard to hit.

Kitazawa, as scheduled, talked about old work and did not even mention the new material in his talk. Chu also talked about old work, but concluded by saying that he had verified the discovery by Bednorz and Müller of a 30-K superconductor. Chu had not taken their work any further than they had: He had not isolated the pure form of Bednorz and Müller's superconductor, nor had he performed a vital test for the Meissner effect (whereby superconductors expel magnetic fields from their interior), a property that—even more than zero resistance, which is difficult to determine accurately—is an unmistakable hallmark of superconductivity. Chu could announce that Bednorz and Müller were right in their assertions, but, like them, he could not say for sure whether this new material really was a superconductor. Nevertheless, that he had so easily reproduced the Bednorz and Müller results was heartening.

Kitazawa was surprised and pleased by Chu's talk; although he was certain that the results from his lab were accurate it was nice to receive independent verification. At the end of the session the chairman, Alex Braginsky, from Westinghouse, who, like everyone else, had heard the rumors from Japan, took the microphone and said, "I believe Dr. Kitazawa from the University of Tokyo has something he would like to tell us." He added, slyly, that Kitazawa would give another talk, at the end of the conference, on Friday. This was news to Kitazawa, who, after checking with Tanaka in Tokyo, agreed to give a short presentation on their replication of the IBM work.

Back in Tokyo the work in Kitazawa's lab went on around the clock. The mixture of barium, lanthanum, copper, and oxygen specified in the IBM paper actually produced a brew consisting of several different phases, or compounds. In Tokyo the researchers had managed to isolate the pure superconducting phase, which turned out to be two parts lanthanum, one part copper, four parts oxygen all doped with a smidgen of barium, $La_{2-x}Ba_xCuO_4$, in the shorthand of chemistry. Kitazawa kept in constant touch by fax, a device well suited to both the hieroglyphics of science and the ideograms of the Japanese language.

On Friday afternoon, December 5, while snow was falling over Boston, Kitazawa told the thirty or forty scientists at the MRS meeting who were interested in superconductivity the latest news from Tokyo. He spoke slowly in strongly accented English. But the pictures, not the words, did the convincing; smudged transparencies of data only hours old. A line traced the vagaries of electrical resistance as the temperature dropped: a gentle, downhill slope until somewhere around 30 K, the tropics of low-temperature physics, where it fell abruptly to zero, as sharply as the edge of a knife. And as the scientists made their way through the snow to Logan Airport, and flew back to their labs all around the world, they began plotting their campaigns. Their old research, which had once so absorbed them, would be put aside. Funds would have to be diverted. It was a whole new ball game.

CHAPTER 2

THE GENTLEMAN OF
ABSOLUTE ZERO

In the winter of 1852, in the village of Kincardine-on-Forth, Scotland, a ten-year-old boy went playing on the thin ice of a local pond. The ice broke, and the boy, whose name was James Dewar, son of a wine merchant, took a brief and, to use a word that had not yet been coined, cryogenic plunge. Dewar, who would one day invent, among many other things, the thermos bottle, was chilled to the bone; he contracted rheumatic fever and was unable to attend school for two years. In his later life Dewar would look back on this accident as fortunate, and on those two years out of school as the most important period of his education. During those two years Dewar developed skills that helped him become one of the greatest experimenters of his day, who brought science into the realm of absolute zero and to the brink of superconductivity.

In those two years Dewar spent a lot of time in the local carpenter shop, where violins were built and repaired. It was a practical education. Before long Dewar was building violins by himself. After he completed his schooling he went to Edinburgh with a violin labeled "James Dewar 1854," proof of his manual dexterity. It was the only résumé he needed.

Apparently Dewar was not interested in enrolling in the University. The well-crafted violin was proof enough of Dewar's mechanical skills to secure him a job as a "laboratory boy" to J. D. Forbes, an authority on the physics of glaciers at the University of Edinburgh. As laboratory boy Dewar served the kind of scientific apprenticeship that even in his day was beginning to go out of style, being replaced by university education.

Dewar's talents were soon noticed. When he was twenty-five he built a mechanical device for representing the arrangement of the six carbon atoms in a molecule of benzene. The elucidation of the structure of benzene was one of the outstanding triumphs of the nineteenth century, which led to modern organic chemistry. The German chemist Friedrich August Kekulé von Stradonitz, or so the story goes, after fruitless wrestling with the problem, had fallen asleep and dreamed of a snake eating its tail. When he awoke he swiftly decoded the dream: The six carbon atoms of benzene were arranged in a ring. Dewar's mechanical apparatus confirmed Kekulé's somnolent-inspired insight, and also showed how the carbons could be linked up in six other structures, which were subsequently discovered by others. Kekulé was so impressed that he asked Dewar, who was not even a university graduate, to come spend the summer in his laboratory in Ghent. Dewar's career was launched.

Back at Edinburgh, as a lecturer and, later, a professor at the Royal Veterinary College, Dewar was astonishingly productive. He investigated the chemistry of chlorine, and the temperature of the sun and of electrical sparks; with Sir Frederick Abel he invented cordite, a smokeless propellant. Most important he became interested in the properties of gases. His activity at Edinburgh was noticed, and in 1875 he was elected to the Jacksonian chair of experimental science at Cambridge, which was originally endowed by Reverend Richard Jackson in 1783 so that its occupant would, among other things, search for a cure for "that opprobrium medicorum called the Gout." This was not a serious constraint; the terms of the chair were general enough to allow Dewar to work on anything that caught his interest. And two years later he was appointed Fullerian professor of chemistry at the Royal Institute, where Sir Humphry Davy and *his* laboratory boy, Michael Faraday, had worked.

Faraday had been a scientist in Dewar's mold, self-taught, thoroughly practical, and an ingenious experimenter. Dewar was said to wander the halls of the Royal Institute late at night, "communing" with the spirits of Faraday and Davy.

While still Davy's assistant, in 1823, Faraday had managed to liquefy chlorine gas. By 1848 he had liquefied all the known gases save oxygen, nitrogen, carbon monoxide, nitrous oxide, and methane. Since these gases resisted his best efforts at liquefaction Faraday called them the "permanent gases," enshrining his failure in chemical jargon. And for years scientists, or, as they were then called, natural philosophers, believed that Faraday's frustrated coinage corresponded to physical reality and that the permanent gases could not be made liquid. Their opinion changed abruptly in 1877, when a young French mining engineer named Louis Cailletet liquefied oxygen and nitrogen (he shares credit with another scientist, Raol Pictet, who also discovered how to liquefy oxygen and nitrogen at about the same time).

Within a year Dewar, who delighted in giving public lectures that "brought the spectator to the frontier of knowledge," demonstrated the fleeting liquefaction of oxygen to an audience at the Royal Institute. Perhaps also recalling his plunge through the ice, Dewar resolved to liquefy the last remaining of the known gases, hydrogen, and to make perhaps the iciest plunge of all, right down to absolute zero. In his mind he kept the image of a truly spectacular public demonstration: hydrogen, "boiling quietly in a test tube." But there was more behind the quest than that.

Absolute zero was to Dewar and his contemporaries what whiteness was to Ahab and the crew of the *Pequod:* an obsessive, almost impossible goal, an ideal of almost horrifying beauty. The goal was first glimpsed by Guillaume Amontons in the middle of the seventeenth century. Amontons, who had lost his hearing in his youth, devoted his life to the careful measurement of temperature. He noticed that at zero degrees Celsius, a convenient standard temperature, since it corresponds to the freezing point of water, the pressure of a cylinder of gas decreases by approximately 1/273rd and that this was true almost regardless of the type of gas. Furthermore, the pressure would drop by the exact same amount with every additional degree Celsius the temperature dropped. He concluded that the pressure of any gas should vanish at about 273 degrees below zero Celsius, the temperature now known as absolute zero.

The industrial revolution of the early nineteenth century turned the attention of many scientists to thermodynamics, the study of heat, largely to aid in the design of more efficient steam engines. The big question was, is there a limit to the efficiency with which a heat engine can convert the burning of its fuel to mechanical energy? The surprising answer was yes, there is, and that it depends on the heat at which the engine is operated. These scientists, too, ran right up against absolute zero. An idealized heat engine operates by exploiting the heat that flows from a hot object to a cold one; from, for example, the steam that fills the cylinder of a steam engine to the surrounding jacket of cooling water. The thermodynamicists showed that only when the lower reservoir was held at absolute zero would the engine be 100 percent efficient.

The mysterious cachet of absolute zero was further increased when the microscopic theory of heat began to be elucidated in the second half of the nineteenth century. Temperature was then thought to be related to the ceaseless motion of atoms and molecules. In a gas at high temperature the atoms rush about wildly, hardly interacting with one another at all. As the gas is cooled this atomic motion slows. As the atoms fly more slowly by each other they begin to feel the attraction of a very weak force, called the van der Waals force after Johannes Diderik van der Waals, the Dutch physicist who postulated it, for which he won the Nobel Prize in physics in 1910. At high temperatures the weak van der Waals force has little effect on the trajectories of the speeding atoms. As the temperature drops the motion of the atoms in the gas slows down, and quite abruptly there comes a point when the atoms all clump together and form a liquid. This sudden change from gas to liquid is an example of what is called a phase transition, the details of which scientists working with high-speed computers cannot calculate precisely even today. Continue to cool it and the liquid undergoes another phase transition and becomes a solid. It freezes, and the atoms become anchored to fixed positions about which they tremble with the remaining heat. The mystery of phase transitions, whereby the forms of matter are transmuted, is captured in these lines by Wallace Stevens:

That the glass would melt in heat,
That the water would freeze in cold,
Shows that this object is merely a state,
One of many, between two poles.

By studying the properties of matter cooled to near absolute zero scientists like Dewar hoped to glimpse the cold geometrical beauty of nature bare, unobscured by the random thermal motion of atoms. Dewar explained the lure of absolute zero in an article on the liquefaction of gases he wrote for the eleventh edition of the *Encyclopaedia Britannica:* "Though 'Ultima thule' may continue to mock the physicist's efforts, he will long find ample scope for his energies in the investigation of the properties of matter at the temperatures placed at his command by liquid air and liquid and solid hydrogen. Indeed, great as is the sentimental interest attached to the liquefaction of these refractory gases, the importance of the achievement lies rather in the fact that it opens out new fields of research and enormously widens the horizon of physical science."

In fact, Dewar's "Ultima thule" is physically unachievable. At exactly absolute zero, according to classical physics, all atomic motion ceases; nature sleeps, isolated from the procession of cause and effect. In the twentieth century scientists have cooled matter to within millionths of a degree of absolute zero, but absolute zero, they know, is impossible. According to quantum theory, matter is really never at rest, but suffers a "zero-point" jitter, a kind of irreducible quantum nervousness.

To equip himself for the assault on absolute zero Dewar needed first of all a test tube that would perfectly insulate its contents from the heat of the lecture hall. Vacuum, he knew, is a nearly perfect insulator of heat. So, in 1892, Dewar fashioned a double-walled glass container, and evacuated the space between the walls. The inside wall of the container was silvered to further insulate the contents against radiant heat. The construction of such a bottle, which today is known to scientists as a Dewar or a cryostat and to picnickers as a thermos, was one of Dewar's triumphs. The slightest imperfection in the thin glass walls and the bottle would shatter when the air was drawn out. The quality of the vacuum needed for adequate insulation was higher than had yet been achieved, so Dewar had to invent new methods of evacuation.

Dewar did all of the delicate glassblowing by himself; he was impatient of slowness and clumsiness in others. To be frank, he was a short, difficult man, with an "artistic temperament," as some of his more generous contemporaries called it. In truth, he was a tyrant, given to rages, and was often cruel. Then, as now, it was common for a laboratory chief to share credit for all experiments performed in his lab, regardless of how much or how little he contributed. Dewar carried this custom to the ungenerous extreme of sometimes failing to give any credit at all to those who largely conceived of and carried out an experiment.

The Dewar flask was perfected in 1892, and given its first public demonstration the following year. The demonstrations were effective in more ways than one. To liquefy hydrogen required funds far greater than were then generally expended on scientific experiments. The Goldsmiths' Company gave Dewar £1,000 to complete his experiments. The *Times* of London noted this, and presciently credited Dewar with the discovery of a new scientific law, the law of increasing expenses.

Dewar finally managed to liquefy hydrogen in 1898. He found it boiled at a temperature of about 20 degrees above absolute zero, or 20 degrees Kelvin (named for the great British physicist Lord Kelvin who, in 1848, had proposed an "absolute scale" of temperature, independent of the properties, such as the freezing point, of any particular kind of matter). By reducing the pressure in the vessel containing the liquid hydrogen he was able to further reduce the temperature to 13 K, still a long way from his goal of absolute zero. Fortunately, fate had given Dewar one more chance. Three years earlier, in 1895, a new "permanent" gas had been discovered, and initial experiments showed that if this one could be liquefied at all it would be at a temperature far lower than that even of hydrogen, which already demonstrated a temperature not much above absolute zero.

During a solar eclipse in 1868 Joseph Norman Lockyer observed what he felt was evidence for a new chemical element, which, to properly reflect its origin, he called helium, from the Greek word for sun, *helios.* Lockyer's interpretation of his solar observation was debated for almost 30 years. The dispute was finally settled with the discovery by Sir William Ramsay of helium among the gases released by the mineral springs at Bath, in Somerset. Ramsay, who received the Nobel Prize in chemistry in 1904, had his laboratory at University College, not far from Dewar's. Dewar needed large quantities of helium to prime his refrigerator for its assault and Ramsay would have been the ideal person to get it for him. But this was impossible.

At a scientific meeting in 1895, Dewar began a feud with the mild-mannered Ramsay, a man whose motto was said to be, "be kind." After the Scots scientist spoke of his desire to liquefy hydrogen Ramsay rose and said that he believed that a scientist in Cracow had already liquefied large quantities. Ramsay was mistaken, and Dewar took the opportunity to attempt to humiliate him. The two men never spoke again.

Dewar was left to collect and purify the Bath helium by himself. On the eve of his experiment one of Dewar's mistreated laboratory boys had accidentally turned a valve and released half of the precious gas to the atmosphere. The remaining helium was still adequate for the attempt, and Dewar proceeded.

Though a fine chemist, Dewar was not Ramsay's equal. The helium gas he had isolated was contaminated with neon. Neon freezes and solidifies at an awkwardly high temperature. Dewar encountered the laboratory version of a homeowner's nightmare: frozen, ruptured pipes. After this failure Dewar abandoned for good his cryogenic research. He finished his days investigating the properties of soap films.

Dewar was among the last of the great artisan-scientists, virtuosi who played nature like a violin. Science was moving into a new, industrial phase. The frontier of knowledge was becoming a place unapproachable by solo expeditions. Teams of explorers with many different skills were needed. To lead their assault came a new breed, the manager-scientist. Among the first of these was Heike Kamerlingh Onnes, the man who would liquefy helium and discover superconductivity.

Born in Groningen, Holland, in 1853, the son of a well-to-do tile manufacturer, Onnes was almost the exact opposite of Dewar in every way. Onnes matriculated at the University of Groningen in 1870 and quickly demonstrated his scientific talents, winning the gold medal of the Faculty of Natural Sciences at Utrecht. This brought him the following year to study in Heidelberg. Returning to Groningen he received his doctorate in 1879 for a thesis entitled "New Proof of the Earth's Rotation." At the end of his defense of his thesis the examiners burst into applause.

At the age of twenty-nine he was appointed professor of experimental physics and meteorology at the University of Leiden. In his inaugural address he laid down his scientific credo, which he suggested be placed over the entrance to every laboratory of physics: *Door meten tot weten,* through measurement to knowledge.

At Leiden Onnes came under the influence of van der Waals. He had become fascinated with van der Waals's law of the continuity of liquids and gases. The law states simply that when a gas is cooled and becomes a liquid the atoms remain unchanged; they are just packed more closely together, held there by interatomic forces. Verification of van der Waals's hypothesis required careful measurements of the properties of gases, carried out at extremely low temperatures. The challenge appealed to Onnes, and to meet it he established his own Cryogenic Laboratory, which was to become world famous.

The success of the Cryogenic Laboratory owed as much to Onnes's organizational talent and inexhaustible patience as it did to his technical prowess. Onnes knew that to carry out the researches he had planned would require the aid of skilled technicians and glassblowers, so in 1901 he established the "Society for the Promotion of the Training of Instrument-makers." This school within his laboratory not only taught technical skills, but introduced its students to the basics of physics so they might better understand the activities of the laboratory. Before long Leiden-trained glassblowers would be essential to laboratories all around the world. And scientists from these laboratories were always welcome to share the facilities at Leiden for their investigations of physics at low temperatures.

The Leiden lab churned out new results at such a furious pace that Onnes published and edited his own journal, *Communications from the Physical Laboratory at Leiden,* which came out regularly and in English. The *Communications* are documents from a transitional time of science and make curious reading. Most scientific papers written today are highly formalized accounts, almost deliberately dull, written to disguise the fact that the research was carried out by fallible, finite humans. Blind alleys down which the authors might have wandered for months are skillfully bypassed. Acknowledgment of colleagues is saved for the footnotes or a few terse lines at the end of the paper ("The authors would like to thank Dr. X, without whose help . . ."). The sheer volume of papers vying for journal space almost makes this terse, pared-down style a necessity. But Onnes was his own publisher and editor, and #108, his account of his liquefaction of helium, was his masterpiece.

In preparing for the assault Onnes and his Leiden colleagues had performed accurate measurements of the relationship among the pressure, volume, and temperature of helium, what is known as the equation of state. With these measurements and a theory due to van der Waals, Onnes was able to determine that helium would liquefy at around 5 or 6 K. This was within the calculated range of his apparatus, so Onnes proceeded to marshal his assault.

Like Dewar, Onnes had to obtain a large quantity of pure helium gas. Luckily, as he explains in #108 of *Communications,* his brother, director of the Office of Commercial Intelligence in Amsterdam, managed to procure for him large quantities of monazite sand, from which helium could be extracted "on favorable terms." So the first hurdle was easily cleared.

The actual liquefaction apparatus had been under construction for years, and had required the skills of all his glassblowers and technicians. For its design it relied heavily on the work of Dewar, and Onnes generously acknowledges this in a paragraph that must only have enraged the irascible scientist. The approach to helium temperatures was managed in a "cascade," the scheme that Pictet had applied in his liquefaction of oxygen. The system involved separate closed cooling systems. In the first oxygen was liquefied. The liquid oxygen then was used to cool the nitrogen system, the liquid nitrogen cooled the hydrogen system, and the hydrogen cooled the helium. Each of these systems was an extremely complex bit of plumbing, made mostly out of glass, with its own gastight valves, compressors, exhaust pumps, gasometers, liquefiers, and cryostat baths. As a colleague would write, "The famous Leiden Laboratory had already come to look like a combination of a physical laboratory with an engineering plant."

Onnes reports that the experiment began at 5:45 A.M. on July 10, 1908. He had already accumulated 75 liters of liquid air the previous day. These were used to liquefy the 20 liters of hydrogen that he had calculated would be needed. "Arrived at this point," Onnes writes, "I resolved to make the reaching of the end of the road at once my purpose."

Now things became tricky, because the slightest amount of air in the apparatus would freeze and, at best, cloud the glass walls of the helium container, obscuring the evidence of his success. By 4:20 P.M. helium circulation was started and the temperature plummeted to new lows. But before any liquid helium was observed the temperature leveled off, and despite all efforts to budge it, it held. By 7:30 they were ready to throw in the towel.

Word that an important experiment was being conducted in Onnes's lab had spread across the University, and throughout the day visitors stopped by. One, a Professor Schreinemakers, suggested that the boiling liquid might be difficult to see. A light was promptly moved to below the container, and there it was, a quiescently boiling liquid as clear and thin as air. They had made 50 cubic centimeters of liquid helium. The temperature was 4.2 K, the temperature of outer space.

Onnes and his assistants tried next to solidify the helium by cooling it further, but it resisted all their tricks. At 10 P.M. Onnes decided that it was finally time to go home. "Not only had the apparatus been strained to the uttermost during this experiment and its preparation, but the utmost had also been demanded from my assistants," Onnes wrote. "But for their perseverance and their ardent devotion every item of the program would never have been attended to with such perfect accuracy as was necessary to render this attack on helium successful." It is easy to see why he was known in the scientific community as "le gentleman du zéro absolu."

With the world's only supply of liquid helium Onnes had the realm of absolute zero all to himself. He embarked on a program of measuring the physical properties of materials in this region where the thermal jig of atoms had been slowed to a near standstill. And of all the properties of matter the easiest to measure was electrical resistance.

Over a century earlier Henry Cavendish had begun to measure electrical resistance, the ability of different materials to carry an electric current, the electrical counterpart of friction. When a current passes through a conductor it loses some of its energy to electrical resistance, and this lost energy takes the form of heat. Incandescent light bulbs and electric toasters work on the principle of resistive heating.

Cavendish wanted to measure the relative resistances of materials in the days before electrical instruments, so he took a direct and painful approach. He had become fascinated by the torpedo fish, a peculiar fish of the type ichthyologists call rays, that had somehow evolved the ability to deliver stinging electrical jolts. In 1773 he began a series of experiments that walked the thin line that sometimes separates science from masochism. Cavendish became, in effect, the first electrical meter. He included himself as part of a circuit connecting the material that was to be tested with a torpedo fish. He would adjust the amount of material to be tested to give him a standard jolt. The method was, at best, subjective. Still, in the hands of Cavendish, an experimenter famed for careful measurement (he had measured the minuscule gravitational attraction between two small balls of lead), it yielded useful results. He found, for example, that he received the same shock across a sample of 5.1 inches of salt water as across a sample of 2540 inches of an iron wire.

In the next 100 years the measurement of electrical resistance became steadily less painful. Once Dewar had invented his flask and was able to perform lengthy experiments at low temperatures, he began to study the dependence of electrical resistance on temperature. He did this work in collaboration with John Fleming, a professor at University College, London, who would later invent the "thermionic valve," or vacuum tube, which was the basis of all pre-transistor electronics. Dewar and Fleming found that the resistance of all metals decreased steadily with decreasing temperatures, and all apparently converged to zero at, or just above, absolute zero. But without liquid helium they were unable to follow up on their observation, and in 1895 abandoned the research. Had they continued they would certainly have discovered superconductivity.

According to the atomic theory of matter, which was already an accepted part of the nineteenth-century picture of the universe, an electric current was just a flow of tiny, electrically charged "billiard balls," known as electrons, through an atomic obstacle course. The electrons would collide from time to time with the atoms and lose energy, various scientists theorized, which was the origin of the

concept of electrical resistance. This fit neatly with Dewar's observations of resistance decreasing with temperature, but there was another possibility. Since the electrons
had been part of the atoms in the first place it was possible
that at a low enough temperature the electrons would
become stuck once more to the atoms and not have
enough energy to shake themselves free. In this case, the
flow of electrons would be halted, and instead of vanishing
the electrical resistance would become infinite at absolute
zero. Now, with an ample supply of liquid helium, Onnes
was ready to put the question to nature, for, in the words
of British physicist C. V. Boys, a contemporary of Onnes,
"an experiment is a question we ask of Nature, who is
always ready to give a correct answer, provided we ask
properly." Onnes was quite unprepared for nature's answer.

Onnes started by reproducing some of Dewar's measurements on platinum and gold, but he quickly discovered that impurities in these metals masked the effect he
was looking for. He found that "even . . . the purest gold of
any mint in the world" was too contaminated for his experiments. Mercury was his next choice, because it could
be easily distilled to extremely high purity. It also had the
advantage that its electrical resistance at the temperature
of liquid hydrogen was relatively large and easy to measure. He anticipated that the resistance of mercury would
be small but finite at helium temperature, but "would fall
to inappreciable values at the lowest temperatures which
I could reach. With this beautiful prospect before me there
was no more question of reckoning with difficulties. They
were overcome and the result of the experiment was as
convincing as could be hoped."

Onnes found that at the lowest temperatures he could reach the resistance of mercury became smaller than he could measure with his sensitive apparatus. On April 28, 1911, he reported to the Netherlands Royal Academy his tentative conclusion that the resistance of mercury became, as near as he could tell, zero at the lowest temperature he could reach, just 2 or 3 degrees above absolute zero. He had even concocted his own theory to fit his data. Onnes stopped just short of claiming total victory and returned to his laboratory.

Another month of experiments and things began to look a little more complicated. Onnes still found that the resistance of mercury vanished at the low end of his temperature range, but when he made a careful plot of resistance versus temperature he found there was something else going on that was very peculiar. At temperatures above about 4.2 K the electrical resistance subsided slowly and steadily with temperature. But then, in a space of only a few hundredths of a degree, the electrical resistance collapsed to zero. It would take another fifty-five years for theorists to explain what Onnes had observed, the abrupt transition of mercury from its normally conducting phase to a phase in which every last vestige of electrical resistance is shed, the superconducting phase.

Onnes could not explain this sudden and total loss of electrical resistance near absolute zero, but he gave it a name, superconductivity, a term that was almost instantly adopted, except for the few who for a while stubbornly insisted on "supra-conductivity." In doing so Onnes began the "super" inflation of twentieth-century science, which has given rise to supernovas, supercomputers, supersymmetry, superstrings, super colliders, and various other such "super" inventions. Superconductors, as Onnes was soon to discover, at least deserve the name.

Onnes found that many other metals became superconducting at temperatures accessible with his apparatus. As far as Onnes's measuring instruments could detect, there was not the slightest electrical resistance in any of his superconductors. Still, he could not unequivocally claim that the resistance was exactly zero; no experiment, then or now, could do that. But Onnes came up with an experiment that came close. He made a superconducting circuit in which he started an electric current flowing. When he removed the battery the current, which in an ordinary wire would become immolated by the friction of electrical resistance, continued to circulate undiminished for as long as he cared to keep the superconductor cold. Over the years many others have reproduced this experiment—the persistent currents have circulated for up to four years, until the scientists found better uses for the liquid helium. The results of these experiments have placed a minuscule upper limit on the resistance of the superconducting state, a number so small that it is quibbling to distinguish it from zero. The current in a superconducting loop would keep circulating in the ring for many millions of times the age of the universe. Perpetual motion, or at least a limited version of it, was possible. Who knew what else might follow? As early as 1913 Onnes was predicting that superconductivity could someday be used to make extremely powerful magnets and other such things far beyond the reach of the technology of his time.

Onnes continued his research until the end of his life, in 1926. In late life "the gentleman of absolute zero" suffered from a bronchial affliction that required him to avoid sharp changes in temperature. He continued to direct his lab from a sunny, warm room in his house in Leiden that looked out on the ships passing on one of the outlets of the Rhine. Despite his early hopes, Onnes would never see a practical application for his discovery.

CHAPTER 3

SCHMUTZ PHYSICS

In the first few decades after its discovery in 1911 by Kamerlingh Onnes, superconductivity seemed to be one of nature's purest poems: beautiful, profound, enigmatic, and evidently without any practical use. At first Kamerlingh Onnes thought that superconductors would be a boon to all sorts of electrical machinery, especially in the construction of extremely powerful electromagnets. He soon discovered, however, that the superconducting state was as frail as it was beautiful, and apparently unsuited for practical applications.

At first glance, superconductors are an engineer's dream come true. Except in the few instances in which the wasted heat is put to good use—heaters, blow driers, toasters—electrical resistance is a nuisance, an expense, and a limitation. The problem becomes particularly clear in the case of that most useful electrical device known as an electromagnet.

When an electric current passes through a wire it generates a magnetic field; the bigger the current, the bigger the field. It's called Ampère's law and, among other things, it makes possible electric motors and most other devices that convert electrical energy to mechanical work. Strong magnetic fields are also used in advanced medical scanners and electron microscopes, and as a means of accelerating elementary particles to the huge energies necessary to investigate the ultimate structure of matter. In theory, Ampère's law gives a simple prescription for producing arbitrarily powerful fields: Keep increasing the current. Unfortunately, there is a practical limitation to this approach. If the coil is wound from a normal conductor it will possess an electrical resistance. The energy lost because of electrical resistance increases rapidly as the current is increased. In fact, it increases as the square of the current, so doubling the current, which doubles the strength of the magnetic field, multiplies the rate of energy loss (and the power bill) by four. Even more troublesome is that lost energy appears as heat, which, in the case of very large magnets, must be removed somehow before the magnet melts or burns. Large magnets are often water cooled, which makes them bulky and expensive. But a superconducting magnet, as Kamerlingh Onnes realized, would lose no energy to resistance; a superconducting electromagnet more powerful than any ever made could be tiny, and powered by the current from a flashlight battery. Then Onnes discovered that there was a catch.

Actually, there were two catches. First, there was a limit to the electric current a superconducting wire would carry before it abruptly regressed to its normal conducting state, the so-called critical current. It was as if a superconducting wire were like a bucket brigade that could deliver water at speeds of up to, say, 10 buckets a minute, but could be pushed no faster. Since magnetic field is proportional to current, the existence of a critical current limited the maximum magnetic field that a coil could generate. This was not a serious limitation, because even for the superconducting materials known to Onnes—tin, lead, and mercury—the limit on the magnetic fields obtainable was still usefully high. The real problem, as Onnes soon discovered, was that the superconducting state was also destroyed by a sufficiently large magnetic field, and not a very large magnetic field at that. When a superconductor is held at exactly the temperature at which it first becomes superconducting, that is, the critical temperature, any magnetic field at all will destroy the superconducting state. The same is true of the critical current. As the temperature of a superconductor is further reduced below the critical temperature it becomes increasingly immune to disruption by magnetic fields or electric currents. Still, for any obtainable temperatures (and remember that the materials known to Onnes only became superconducting at a few degrees above absolute zero; they could not *be* much colder) the size of the magnetic field that would destroy superconductivity was minuscule. At least for the time being, superconductivity was a scientific curiosity and nothing more.

But as a scientific curiosity it attracted the interest of physicists all around the world. As more labs obtained the equipment necessary to do low-temperature physics—the helium liquefiers, the Dewars and such—more groups began to experiment with superconductivity. Still, in the first twenty years after its discovery only about ten papers about superconductivity were published. By 1933 eleven superconducting elements and fifty superconducting alloys and compounds had been found, but beyond this little new was learned about superconductivity itself. The results of many experiments seemed to contradict one another, and it took a surprisingly long time before scientists realized that the fault was not nature's, but theirs. This was not the last time in the story of superconductivity scientists would discover themselves blinded by their own preconceptions.

The biggest problem was that, until 1933 and the pioneering work of Walther Meissner and R. Ochsenfeld, physicists tacitly had assumed that a superconductor was essentially a normal conductor that for some reason failed to manifest any electrical resistance. Meissner and Ochsenfeld showed that this assumption, while reasonable, was incomplete. A superconductor was also a perfect magnetic shield, that is, it cannot be penetrated by a magnetic field, a completely independent property. Reinterpreted in terms of the Meissner effect, experiments that had seemed to be contradictory were shown to be perfectly consistent with one another. Superconductivity was suddenly no longer paradoxical, it was merely inexplicable.

In the thirties researchers began to think of applications of superconductors that did not require either high currents or large magnetic fields. Their strategy was to exploit the apparent weakness of superconductors. Superconductors are sensitive to heat? Very well, use them as heat, or infrared, detectors. During the Second World War superconducting bolometers, as they were called, were developed to detect the movements of warm-blooded troops through the darkest of nights. The war ended before they could be put to use.

Superconducting microwave circuitry, computers, and frictionless bearings were designed, but, largely due to finicky abhorrence of magnetic fields and the extremely low operating temperatures required, few applications made it off the drawing board. Superconductivity became a backwater of science, a phenomenon unexplained by physicists and unused by engineers.

Shortly after his arrival at the University of Chicago in 1949 John Hulm, a young British physicist with a brand-new Ph.D. from Cambridge's Mond Laboratory, was invited to a dinner party at the home of the head of his department, Andrew Lawson. Hulm, who is now chief scientist at the Westinghouse Research & Development Center in Pittsburgh, had come to the United States to get what scientists of his generation in England half-seriously called the BTA degree—Been To America—an essential credential to be obtained before settling down in British academe. This was Hulm's intention, too, but he remained in the United States.

As a graduate student at the Mond, Hulm became interested in an odd material called barium titanate, a pale yellow, transparent crystal. Barium titanate is one of an interesting class of materials known as ferroelectrics, which are, loosely speaking, the electrical equivalent of permanent magnets. Below a certain critical temperature, around the boiling point of water, one side of a barium titanate crystal spontaneously develops a positive electric charge while the other side develops a negative charge. To better understand ferroelectrics Hulm required large crystals of barium titanate, which in those days could not be purchased from a chemical supply house. The only way for Hulm to get the crystals was to grow them himself, and this proved to be a difficult task for Hulm, who at the time was not much of a chemist. Fortunately, he was saved by the academic grapevine.

Hulm's professor at the Mond was Sir Lawrence Bragg, a Nobel laureate who had invented the technique of determining the atomic structure of crystals with X rays; Bragg was a close friend of Paul Scherrer, a professor at the Swiss Federal Institute of Technology, the ETH in Zürich, who would later teach Alex Müller; Scherrer had a brilliant young student from Frankfurt am Main, Germany, named Bernd Matthias; Matthias had invented a good method for growing barium titanate crystals. Matthias told Scherrer about the technique, Scherrer told Bragg, and Bragg told Hulm.

Using Matthias's technique Hulm grew some barium titanate crystals, did experiments on them, and published the results in *Nature*. When Matthias saw Hulm's *Nature* article he began to seethe. Matthias's method was a technical innovation, a tool, and not a scientific discovery. The Cambridge physicists had asked and received from Scherrer permission to freely use Matthias's work. Unfortunately, nobody had bothered to ask Matthias. Hulm had not committed a scientific impropriety, but a courtesy had been overlooked, and from Matthias's point of view, Hulm was a thief, plain and simple—a despicable and dishonorable scientist. Matthias wanted revenge, but for the time being he settled for writing a long, angry letter to Hulm, though he didn't forget Hulm's perfidy.

Says Hulm of the dinner party at the Lawsons' that took place almost forty years ago, "I can envisage the scene just as clearly as if it had happened yesterday." The Lawsons had a beautiful house, he recalls, which was filled with faculty members and their wives, "all dressed to the nines and drinking cocktails." Early in the evening Lawson introduced Hulm to someone he thought he should meet, Bernd Matthias.

Matthias was thirty-one years old, a dark and handsome man, about the same size as Hulm and several years older. When he got angry or excited, which happened frequently, he spoke in italics; at least this is the way Hulm remembered him years later. After Lawson's introduction of Hulm, Matthias barked, *"How do you spell your name!"*

Hulm spelled his name.

"Were you at Cambridge?"

"Yes."

"Are you the guy that stole my work on barium titanate?"

"I didn't steal your work. We had a visit from your boss, Scherrer . . . "

"Damn liar! I challenge you to a duel."

Matthias had a flair for the bombastic, but he was angry. Employing the diplomatic skills that in later life would serve him well on a two-year stint as scientific attaché to the U.S. embassy in London, Hulm deflected Matthias's challenge.

"I'll fight you," Hulm said calmly, "if that's what you want—later. Right now Mary Ann Lawson is about to serve dinner. It will cause an incident and spoil the dinner. You don't want to spoil your hostess's party, do you? Why don't we wait until we've eaten dinner and then we'll settle it."

"All right, but I won't forget!"

Fortunately, Matthias was not seated near Hulm during dinner. But as soon as dinner was over he cornered Hulm, and in the same fierce tone said, *"So you know something about ferroelectrics, huh?"* It turned out that Hulm was the only person in the Metals Institute at the University of Chicago who knew anything about ferroelectrics, a field in which Matthias was a world expert, and Matthias was starved for scientific conversation. Hulm admitted that he knew a little bit about ferroelectrics, and that, yes, he was familiar in a general sort of way with Matthias's latest results on potassium tantalate and the niobates. *"Let me tell you . . . "*

The two young scientists spent the rest of the evening sitting on the floor against the wall talking about physics, planning an entire program of research. Matthias erupted with ideas; they flowed from his fertile mind in a continuous stream that in later life would keep an army of graduate students and colleagues constantly employed. By the time the last couples were saying their goodbyes the two scientists had agreed to meet the next morning, a Saturday, and get to work.

"It was a roller coaster," Hulm recalls. "It was drop everything else. I had a number of other experiments I was in the middle of. He was such an exhilarating guy to work with that you just couldn't resist. He was very dominating, he had so many ideas . . . " For the first six months Matthias and Hulm were completely absorbed in measuring the properties of ferroelectrics in the vicinity of absolute zero. The low-temperature laboratory at the University of Chicago was located in the west stands of Stagg Field, just down the hall from where Enrico Fermi had built the world's first atomic reactor during the war. The stands were unheated, and during the brutal Chicago winters Fermi's scientists had kept fashionably warm by bundling in raccoon coats that had been abandoned after intercollegiate football was banned on the Chicago campus. The low-temperature physicists had to make due with less stylish accoutrements. Despite all this Hulm and Matthias found the west stands to be a nearly ideal place for a laboratory. Physicists like to change things, and nobody minded what they did to the west stands. They could drill holes through walls, floors, ceilings, run pipes and cables to their hearts' content. "The trouble with modern laboratories in fancy buildings," Hulm says, is that "you can't make a hole in the wall without making people angry."

One day Hulm gave a seminar on their latest work in ferroelectricity. Fermi was in the audience, and after Hulm was done he remarked that while ferroelectrics were very interesting, and perhaps not perfectly understood, people had a pretty good general idea of what was going on inside them at the microscopic, quantum-mechanical level. Fermi didn't understand, therefore, why Hulm and Matthias were paying so much attention to ferroelectrics and ignoring what seemed to him to be a much more fundamental problem—discovering the mechanism of superconductivity.

Matthias and Hulm were impressed by Fermi's remarks, and both decided overnight to turn their attention to superconductors. The question was what approach to take. Hulm was in favor of an orthodox physics approach. "What we ought to do," he said, "is take a simple superconductor like tin and grow a single crystal out of it and measure some properties. That's the way the Cambridge people looked at the world. By the way, that is a very meaningful approach to life, you can learn a lot..." Perhaps, but that was not what Matthias had in mind.

Hulm's approach was the one any physicist would have chosen. But as far as Matthias was concerned, it was a case of examining the trees and ignoring the forest. He proposed a chemical approach, a broad search for new superconductors, preferably for ones with higher critical temperatures. After assembling the forest of superconductors Matthias believed that it would be possible to discern its borders and the familial traits of its trees. Years later Matthias explained the philosophy of his approach to an audience of physics students: "Let's look at so many instances of one given phenomenon that at least we can get a... feeling for what the crucial conditions are. If we do this, then relying on the correctness of these conditions, we can make predictions. This is what I did in superconductivity, and formerly in ferroelectricity. And the fact that these compounds [that were discovered] become superconducting is—in a way—a justification for this approach."

"I didn't buy it at first, because I came more from a physicist's viewpoint," Hulm recalls. "But inside a year I realized it was a fruitful approach and we had it virtually to ourselves."

Of all the tools available to materials scientists, the most indispensable is probably a piece of paper known as the periodic table of the elements. Late in the nineteenth century the Russian chemist Dimitri Mendeleev noticed that when the known elements were arranged in order of increasing atomic number, that is, the number of electrons that swarm around their nuclei, their chemical and physical properties recurred periodically. He made a chart of the elements that was arranged so that the periods fell neatly into columns. In order to do this, Mendeleev found that he had to leave certain gaps, holes in which he believed would fit elements yet to be discovered. Since then all Mendeleev's holes have been filled, and the observed periodicity has been explained by the quantum theory. The electrons in an atom are organized in a series of concentric shells, each of which can only contain a certain, small number of electrons. The chemical properties of an atom are determined by the number of electrons in its outer shell; atoms with the same number of outer electrons, although they will differ in weight and size, will have similar chemical properties. Elements with the same number of outer electrons occur in the same column of the periodic table. The periodic table is a concise and orderly guide to the properties of the chemical elements. To Matthias, who has been called Mendeleev's modern successor, the periodic table was the first place to look for new superconductors.

It was obvious to Matthias and Hulm that the place to concentrate their search was an area of the periodic table known as the transition metals, which include niobium, tungsten, titanium, and zirconium. Transition metals are extremely hard and have high melting points, which make them very difficult to work with. Before the war investigators looking for new superconductors made compounds and alloys of the soft metals, such as tin, lead, and indium, but the transition metals proved too difficult for most of them. Only Walther Meissner and his colleagues in Berlin did any work on transition-metal compounds, because Meissner had close ties to the German tool steel manufacturers who would send him samples of the compounds they cooked up. Meissner's work indicated that the transition metals should provide a rich hunting ground for new superconductors; Matthias was planning a safari.

From the start transition metals proved to be an experimentalist's nightmare. The method they used in making samples was dirty, tedious, and unreliable. They would take the raw ingredients—niobium, carbon, whatever—grind them into a fine powder, and sinter the powder in an oven, that is, heat it until all the granules softened and fused together. An added complication was that the cooking had to be done in a vacuum because in the presence of the smallest amount of oxygen the compound would become oxidized, and in those days before Bednorz and Müller oxides were not considered to be likely candidates for superconductors. With the vacuum pumps of the 1950s there was always a little bit of oxygen left in the container, and so their samples were at least partially contaminated with oxides. "It was agonizing," Hulm recalls. "I thought, what the hell had we gotten into? We couldn't handle the damn stuff . . . we sweated over this, we made a lot of compounds and alloys in that way for two years."

In those two frustrating years, despite the difficulties, Matthias and Hulm discovered a number of new superconductors and began to formulate some general, empirical rules about which compounds would be superconductors and which would not. Then Matthias decided to take a job at Bell Labs. Even with a long list of publications and an international reputation he had not been made an associate professor at the University of Chicago. "He wasn't tactful with people he didn't think were his equal scientifically," Hulm tactfully explains. Matthias made it no secret that he didn't think many of the professors were as good as he, and that he would not be judged by such men. It may have been the truth, but even (or especially) on a university campus truth is not always welcome.

Hulm, who was made assistant professor, remained in Chicago and continued the search for new superconductors. He and Matthias still kept in constant touch but the distance between Bell Labs in New Jersey and Chicago protected Hulm somewhat from the typhoon force that was Matthias's imagination.

Hulm was never fond of the tedious sintering process,

and gradually became convinced that there had to be a better way of making new materials. One day, while leafing through a metals journal, he came across an article that intrigued him. It described a new type of furnace that was capable of achieving extremely high temperatures, high enough to instantly meld his intransigent transition metals. The heating was done by an electric arc, a fat, fiery bow between two tungsten electrodes. The furnace had a hearth made of copper, and it and the electrodes were cooled by water. Hulm immediately summoned one of his graduate students, a brilliant young chemist named George Hardy, and within three months the two of them, with the aid of a machinist at the Metals Institute, designed and built an electric arc furnace.

"It was unbelievable," Hulm recalls. "It was like emerging into daylight after long darkness." Suddenly they were able to fuse refractory metals into compounds and alloys that would have been unthinkable by the sintering process. The furnace worked so rapidly that oxidation was no longer a concern. He and Hardy immediately set to work with their new magic hearth. And just this once Hulm decided not to tell his best friend and frequent collaborator, Bernd Matthias. Having found the route to a new land he wanted to wander around on his own for a while and enjoy a few moments of serenity.

Without the guidance of theory, searching for new superconductors is a lot like exploring a country without a map. Where are the rivers? Where are the mountains? The only guide the researchers had was their intuition. They knew in a few cases which materials were superconductors and which weren't, and from this knowledge they formed tentative rules. Some of the rules were conscious, rational, and easily expressed. For example, it was pretty clear to scientists in the early 1950s that magnetism and superconductivity were mortal enemies. Atoms that had a magnetic moment, like iron or manganese, were therefore not superconductors in themselves and would probably destroy superconductivity if incorporated into a compound or an alloy. But there were other, preconscious, ineffable rules that were formed by a lifelong intimacy with the periodic table and the strange logic of quantum mechanics. Such rules were Matthias's particular strength. According to Stanford University's Ted Geballe, who next to Hulm was Matthias's closest friend and collaborator, Matthias had "a way of thinking about things that was from his own point of view rational, but it wasn't necessarily purely scientific. Something would remind him of something. He had a way of going through the infinity of parameter space, keeping it in his mind. He had a mental model of his own which he used to very good advantage."

With the electric arc furnace Hulm and Hardy made a rapid tour of the periodic table. They started by making compounds containing transition metals and nitrogen and carbon, the same materials that Walther Meissner had explored in Berlin. They found that Meissner had made a few small errors, and they corrected these and pushed on. The logical thing to do was to survey systematically the first long series of elements in the periodic table: lithium, beryllium, boron, carbon, nitrogen, oxygen. Which is what they did, omitting beryllium because of its reputation for toxicity. Beryllium is used in making some of the phosphors on a television screen, and Hulm knew of several people who had died handling it. As they traversed the table they made several interesting discoveries, including the first oxide superconductor, niobium monoxide. After completing the first long period they moved on to the second, which starts with silicon, an element with four valence (outer shell) electrons that falls just below carbon on the periodic table.

They started by investigating compounds containing silicon and vanadium. What emerged from their furnace depended critically on the initial proportion of vanadium to silicon in the mix. Even with two elements the number of different ways in which the atoms could come together to form a solid, what chemists call phases, was essentially infinite. To make matters worse, many different phases would often form in the same sample. In short, the vanadium-silicon system was a mess. Only by dint of dogged work were Hulm and Hardy able to make sense out of it. Today, scientists in their position would have recourse to a whole armory of analytic tools—X-ray diffractometers, spectrometers, scanning electron microscopes, and much more—to help them unravel the structure and composition of their samples. Hulm and Hardy were almost flying blind. The best probe they had was superconductivity, because if only 1 percent of their sample was superconducting, that 1 percent, by virtue of the Meissner effect, would be impervious to a magnetic field and would show up as clearly as a bullet lodged in the body does on an X-ray film.

Hulm and Hardy made scores of samples before hitting in 1954 on the magic ratio, three atoms of vanadium for every atom of silicon, V_3Si. The resulting compound was a superconductor with an extraordinarily high transition temperature, 17 K, which in those days was near the record. Moreover, crystallographic analysis showed that the atoms of V$_3$Si were arranged in a structure never before seen among superconductors, a structure which now goes by the name A15. Hulm and Hardy wrote a paper and sent it off to *Physical Review Letters.* But Hulm still didn't tell Matthias anything. He wanted to save that surprise for the next time they met face-to-face.

The opportunity soon came at a meeting at the General Electric Research Laboratories in Schenectady, New York. The two friends met and headed right for the bar to have a drink and swap stories. Matthias went first, telling Hulm of a new high-temperature superconductor he had discovered at Bell. Then it was Hulm's turn, and as he recalled thirty-five years later, Matthias was "fit to be tied!"

Matthias immediately asked how Hulm had managed to make this new material. Not by the old sintering method, said Hulm, who told him about the electric arc furnace. *"That's no good,"* Matthias snapped back. *"You've got tungsten electrodes, a copper hearth. Those things melt at two thousand Celsius! Your compound is probably full of tungsten and copper! You've probably got a tungsten alloy in there!"*

"Bernd, you'd think so, but it's just not true. With the water-cooled hearth the tungsten doesn't get hot enough to melt and neither does the copper. I've run analytical tests . . . ."

"I don't believe it," Matthias replied. The two spent the meeting talking, and Matthias was fascinated. He asked Hulm to send him the plans for his furnace. Hulm had held out on Matthias long enough to savor his triumph, but "being his friend and ally, as soon as I got back to Chicago I sent him the plans." When the plans arrived Matthias sent them right down to the Bell Laboratories machine shop,

which quickly produced a duplicate of Hulm's furnace. In a very short time, after satisfying himself that the furnace didn't contaminate the samples it produced, he was in the arc furnace business.

For Matthias the arc furnace became *the* tool. With it he could make in ten minutes compounds that would previously have taken him days. "It's funny how a trivial thing like the arc furnace can influence people's lives and what they do. We didn't invent the arc furnace, it was an article by a guy from a small start-up company that was using it for something totally different, nothing to do with superconductors," Hulm says.

With his new tool the first order of business for Matthias was to explore the new world Hulm had discovered. What other combinations of elements would crystallize in the A15 structure? He had only to look at the periodic table for elements similar to vanadium and silicon and combine them in the magic ratio, 3 to 1, that Hulm and Hardy had uncovered. Before long he had discovered a new A15 made of niobium and tin, and this one had a transition temperature even higher than that of V_3Si—above 18 K— which was confirmed by Ted Geballe, who was just beginning a lifelong collaboration with Matthias.

Matthias and his collaborators discovered hundreds of new superconductors and postulated a number of rules of thumb that, in many cases, could be used to predict whether or not a material would become superconducting and, if so, at what critical temperature. He began splitting his time between Bell Labs and the University of California, San Diego; he had collaborators and students everywhere. Matthias was an immensely charming man, cultured, fluent in many languages. He had the uncanny knack of creating an almost instantaneous rapport with whomever he met, from bellboy to corporate president. He knew his own worth and was not afraid to broadcast it. In the early 1960s, before he settled on UC San Diego, he had many generous offers from top universities and corporations. At the peak of this bidding war Matthias attended an American Physical Society meeting in New York. One day he hopped into a taxi and the driver asked, "Where to?"

"Anywhere," Matthias replied, "they all want me."

Matthias enjoyed a good argument; life to him was boring if he was not treading, if only lightly, on someone's toes. The toes that he most liked to tread on, and not lightly, were those of the theorists.

Even after 1957, the year John Bardeen, Leon Cooper, and Robert Schrieffer published their epochal theory of superconductivity, theorists were vulnerable to Matthias's spirited attacks. The fact was that no theorist had actually predicted a new superconductor, let alone its transition temperature. To Matthias this was a damning truth. He loved to lecture, and was brilliant at it, always talking off the top of his head. He said whatever was on his mind, and one of his favorite topics was the theoretical predictions —every one of them wrong—that litter the literature of his (and, in fact, all) science. A talk he gave in 1971 is representative.

Theorists and their predictions, he said, "clutter up the literature, they confuse the mind, and they give all of us a bad image. Because if they predict, basing it [that prediction] on something, and then fail, we have only two choices. Either they [theorists] are stupid, which they aren't, or they predicted on the basis of something that isn't true. Now which of the two choices would you choose?"

Not surprisingly, Matthias was not universally loved by

all physicists. They called his empirical approach alchemy. The word conjured up the image of a crackpot working alone in a basement dumping newt's eyes and bat's wings into a boiling cauldron. The Germans referred to his work as *schmutz* physics, dirty physics. Matthias didn't mind. He believed in the empirical approach and would say, If you don't understand it, that's too bad for you. You do your physics and I'll do mine.

Some of Matthias's skepticism toward theory was probably justified. As Meissner's work demonstrated, scientists are often prone to a curious blindness, sometimes accepting offhand speculation as fact. In the 1950s there was another, equally unfounded, belief about superconductors: Really large magnetic fields were impossible to generate with superconductors because either the critical current or the critical magnetic field, that is, the field sufficient to extinguish superconductivity, or both, were always too low, and this was a real physical constraint, built into the universe, not just an artifact of poor experimental technique.

The evidence for this belief was, at best, spotty and anecdotal. Until the 1930s no known superconductor could survive in much of a magnetic field, nor could they carry a great deal of current; they were thus disqualified as technological miracles on two counts. Then a new alloy, a combination of bismuth and lead, was found to be a superconductor with a reasonably high critical magnetic field, one that would make it quite useful except that it had a truly feeble critical current. It could carry only the merest trickle of electricity before losing its superconductivity and returning to normal, and so, even though bismuth-lead alloys could withstand large magnetic fields, they could not be used to generate them. Before long Kurt Mendelssohn, an Oxford theoretician, came up with a very plausible theory of why high currents were not compatible with high fields. (His theory also happened to be wrong, but it took decades before anyone noticed.)

Mendelssohn's idea was that the bulk of the bismuth alloy was not, in fact, superconducting. On the microscopic level the alloy, he believed, consisted of tiny, normal grains, connected by very fine, superconducting filaments. The narrow filaments could no more carry a substantial current than a soda straw could suck up the Amazon. The explanation was so reasonable that nobody was particularly bothered that these filaments had not been observed. After all, they were supposed to be very fine.

One day in 1960 Matthias was having lunch with Rudi Kompfner, a friend of his at Bell Labs. "Why don't you do something useful for the Laboratories," Kompfner asked Matthias. Kompfner was trying to build a compact maser, the microwave forerunner of the laser that is used for, among other things, satellite communication. To minimize thermal noise—the whoosh and crackle between radio stations caused by the incessant jittering atoms—which can mask weak signals, masers were already operated at liquid-helium temperatures. Masers also need magnetic fields, large ones, but within the range of then-current superconductors. These fields were also quite feasible with ordinary magnets, but superconducting magnets would be smaller, lighter, and far more energy efficient. Since the maser had to be cooled to liquid-helium temperatures in any case, the use of superconducting magnets seemed a logical choice. Kompfner had seen a small maser with a superconducting magnet at MIT's Lincoln Laboratories. It was wound with niobium wire and it worked, but just barely. Kompfner asked Matthias if he could do any better. Matthias assured him that it would be no problem at all. "At that time," Matthias would write, "I promised him at least 12,000 gauss." For comparison, the strength of Earth's magnetic field is about one-half of a gauss.

Matthias suggested that Kompfner use an alloy discovered by Hulm, a mixture of the metals molybdenum and rhenium. It was the only superconductor malleable enough to be made into wire that also had a reasonably high transition temperature, that is, the temperature at which a substance first becomes superconducting. Kompfner organized a small group of scientists from Bell Labs' Department of Metallurgy who quickly built a magnet that met almost all of Matthias's predictions, except that it actually generated a field of 16,000 gauss. Nobody complained.

In the wake of this success Matthias suggested off the top of his head three other high-temperature materials: the niobium-tin A15 compound he had discovered with Geballe, niobium-zirconium, and niobium-titanium.

The A15 compound of niobium and tin, which has the chemical formula Nb_3Sn, was the most promising candidate. The only problem was that it was very brittle, so it would be hard to wind wires made from it around forms to make magnets. But the Bell Labs researchers came up with an ingenious solution. They started by taking a long, hollow tube of pure niobium metal, and stuffed it, like a cannoli, with a powder of niobium and tin. This confection they then drew down, extruding it through a series of holes of decreasing size, until it reached the desired thickness. From an initial tube 5/8th of an inch thick and 20 feet long they were able to make from 10,000 to 20,000 feet of wire. At this point the wire was not yet superconducting; the niobium and tin had not yet chemically combined. The wire, which was still quite ductile, would first be wrapped into the form of a magnet. Then the whole thing would be baked in a furnace set to 1000 degrees Celsius until the niobium and tin melted and mixed.

It fell to a Bell Labs scientist named Eugene Kunzler to test the Nb_3Sn wire. At the start of this project Kunzler made a bet with his boss, Morry Tanenbaum. Kunzler proposed that he receive a bottle of scotch for every thousand gauss above 25,000 that his wire achieved. The two of them haggled for a while then finally agreed: Kunzler would receive a bottle of scotch for every 3 kilogauss over 25,000 if he would buy Tanenbaum a Beefeater martini for every additional week it took Kunzler to complete a long-overdue manuscript he was supposedly writing for Tanenbaum. To avoid bankruptcy, not to mention cirrhosis of the liver, they agreed to a limit of 100 kilogauss and 50 weeks.

On December 14, 1960, Kunzler made his first measurement of the critical field of a length of Nb_3Sn wire. The wire was placed in the field of the strongest magnet available at Bell Labs, an 88 kilogauss copper solenoid. Cranked to its limit this giant magnetic field could not destroy the wire's superconductivity. According to the formula, Tanenbaum owed Kunzler twenty-one bottles of scotch, which Tanenbaum happily paid. Even more amazing was the fact that at these enormous fields the wire was still carrying a huge current, fifty times larger than had been observed in bulk samples of the material. Ted Geballe threw a party to celebrate the discovery. At the party Bell's patent lawyer asked Matthias what materials he had originally suggested to Kompfner. Matthias wrote them down on a matchbook, which later became an important document in the battle over the superconductivity patents.

With this inarguable evidence that superconductors could tolerate large magnetic fields while carrying substantial currents the myth that these two could not coexist was shattered. With this new outlook, suddenly a lot of puzzling experiments that had more or less been ignored made sense. A theory that would account for the ability of some superconductors to carry enormous currents was contained in a paper that had been written in 1955 by a Russian physicist, A. A. Abrikosov, and quickly forgotten. Critical current and critical field were independent of one another. The size of the critical field depended on the material and could in principle be very large. The amount of current a material could carry depended largely on microscopic defects in the material upon which magnetic-field lines could become snagged in a manner similar to the way leaves and twigs floating down a river catch on rocks and branches dipped in the river's flow. The snagging along the way prevents the leaves and twigs from gathering in one place and forming a dam. In just this way microscopic defects prevent the magnetic-field lines from clumping together and strangling the flow of electric current.

After Kunzler's discovery superconductivity finally began to fulfill its promise fifty years after its discovery. Once the news got out people at different industrial and academic labs raced to make the first supermagnets. Before long superconducting magnets became common laboratory tools. Many of the recent discoveries about the ultimate nature of matter and the origin of the universe were made possible by huge atom smashers and detectors that only superconductivity made possible. Plans for magnetically levitating trains, generators, motors, bearings, transmission lines, energy storage devices, and fusion reactors began to make it off the drawing board as promising prototypes. But as intriguing as these new devices were, they were not going to change the world, not yet. Liquid helium, which was required to cool all known superconductors, was just too damn cold.

CHAPTER 4

THE OUTSIDERS

The meeting of two personalities is like the contact of two chemical substances: if there is any reaction, both are transformed.

—Carl Jung

Over the decades governments and large corporations have learned that it is wise to fund not only research with direct and immediate applications, but also fundamental research with no obvious practical use at all. Among the largest corporate sponsors of fundamental research is the computer giant IBM, which spends about $3 billion annually on research and development, much of which is not directly linked to improving IBM's product line. This research is carried out by over 3000 scientists at three locations worldwide. The smallest of these labs by far, with only about 200 scientists, is the IBM Zürich Research Laboratory, built atop a gentle hill in the suburb of Rüschlikon.

Among the scientists in 1983 at IBM Zürich was K. Alex Müller, who was born in 1926, the year Kamerlingh Onnes died. Müller is an irascible Swiss physicist with a wiry gray beard, unruly hair, and bushy eyebrows. At parties his stock of small talk is quickly exhausted; his mind is never far from his research, or his other love, philosophy. In some things Müller has been a late starter. When he was fifty a friend remarked, "The problem with you, Müller, is that you have no minor vices." He immediately took up smoking and collecting pipes. And at approximately the same age he became interested for the first time in superconductivity.

Müller's interest in science had begun in high school. After the Second World War surplus radio parts from military equipment discarded by the Allies were easy to come by, and Müller began to hack around with them, building receivers and transmitters. It is a story common among European physicists of Müller's generation, many of whom share his fond recollection of American 6L6 pentodes and push-pull amplifiers. Spies were also common in postwar Switzerland. One day Müller was in his attic, banging out Morse code with a transmitter of his own construction, when the police arrived. He was promptly arrested for espionage, but fortunately was able to convince the authorities that he was, as he says, "a reasonable guy," and they let him go.

In order to better understand what made his radios work Müller decided to study electrical engineering. Fortunately, his high school chemistry teacher persuaded him instead to pursue physics. So he enrolled as a physics major at the Swiss Federal Institute of Technology, the ETH, reasoning that "if I was not satisfied, I could still fall back and have an electrical engineering job if I wanted."

The atomic bomb had just been unleashed on an unsuspecting world, and so when Müller arrived in Zürich he became part of what he calls the "atomic bomb semester." He found that "instead of having the usual ten guys who later would become high school teachers, there were forty-six of us, all interested in nuclear energy." To reduce the size of the unwieldy class the professors were especially tough on their students. Müller thrived on the challenge, and before long he knew that he would never become an electrical engineer. He had the ineffable quality —call it hubris—that makes great physicists. He thought he could beat nature at its own game.

For most of his life Müller has relied on his strong intuitions about nature, gut feelings that he feels absolutely compelled to test. "This is something which is very important for me," he says. "If I would think of something, prepare a theory and it doesn't work, I become sick from it. Really sick, physically sick—it's very close." It is not surprising, then, that Müller had been drawn to the works of his countryman, Carl Jung. Every three or four months he gets together with his brother, a psychologist, to discuss the consciousness, the human spirit, and other such numinous concepts. "If I ever stop being able to do useful science I would like to devote myself to these things full time," he says.

Jung wrote what could be Müller's scientific credo: "We should not pretend to understand the world only by intellect; we apprehend it just as much by feeling. Therefore the judgment of the intellect is, at best, only the half of truth, and must, if it be honest, also come to an understanding of its inadequacy." The other half of truth Müller finds in his dreams, to which he attends very carefully. When he was in his mid-twenties he began to write them down, and by now he has collected several thousand pages of them. Sometimes the meaning of a dream will evade Müller's waking mind for many years. But often understanding dawns more rapidly. In one of his frequent dreams Müller is giving a scientific talk to a large, restless audience. As he speaks his listeners, one by one, drift out of the room. Once awake he says the meaning is immediately clear: "Alex, you are doing it wrong! It is the interplay between the unconscious and the conscious. Of course, at that moment I make an intellectual switch, and start examining what I'm doing with the eyes of somebody who doesn't like me. I get really critical of what I'm doing, and look at all sorts of things I had previously ignored. And then I find it, of course. The dream cannot tell me where I went wrong. What it tells me is more general."

Navigating by his dreams, Müller obtained a doctorate in physics from the ETH in 1958, spent five years as a project manager with Battelle Institute in Geneva, and finally ended up at IBM Zürich. His field of interest was, broadly speaking, solid state physics, the study of the bulk properties of matter. More particularly he was interested in ferroelectrics, materials in which an electric field can become locked in much the same way as a magnetic field is locked in a permanent magnet. IBM had investigated ferroelectrics as a possible basis for a computer memory, and finally abandoned them when research showed they could not outperform existing technologies. Nevertheless, Müller found them fascinating and was determined to work on them.

In his next two decades at IBM Müller gained an international reputation for his work in ferroelectrics and, more particularly, on phenomena known as structural phase transitions. These phenomena are related to the central mystery of solid state physics: How does the repetitive, three-dimensional arrangement of the atoms in a solid determine its physical properties (such as color, strength, and electrical resistance)? The soft, black graphite in a pencil and the hard, translucent diamond are both made of carbon atoms; the costly difference is in the way the carbon atoms are stacked, that is, the crystal structure. Under the extreme conditions of temperature and pressure found deep within the mantle of the earth the atoms in graphite spontaneously rearrange themselves to form a diamond. Diamond and graphite are two of the solid phases of carbon, and the transition from one to the other is known as a structural phase transition.

Phase transitions are one of the unifying concepts of solid state physics. They crop up everywhere. The drastic change in the physical properties of a metal when, cooled to nearly absolute zero, it changes from a normal conductor to a superconductor is also a phase transition, not of the crystal lattice, but of the sea of conducting electrons. In a sense, Müller himself underwent a kind of phase transition when, at an age when most physicists slow down the pace of their research and move on to teaching and managing the research of others, he abruptly changed fields, concentrating his research energies on superconductors. Müller had been a manager; he liked it and, with his strong psychological intuition, he was very good at it. His greatest triumph was hiring a young physicist named Heinrich Rohrer, and then teaming him up with a physicist named Gerd Binnig. Binnig was working on a radically new instrument that he hoped would someday produce images of individual atoms on the surface of any material. In 1986 Binnig and Rohrer shared the Nobel Prize in physics for their invention of the scanning tunneling microscope. "I really felt satisfied as a manager," Müller says. Successful as a scientist, manager, and teacher, Müller was filled with the desire to become a student once more. It was an unusual path for a man of his experience to travel, but in 1980 a random confluence of chance and circumstance landed him on the path of no resistance.

John Armstrong, now IBM's director of research and an old friend of Müller's, had for years been trying to get Müller to spend some time visiting the Thomas J. Watson Research Center, the largest of the IBM labs, located thirty miles north of New York City in Yorktown. As long as his children were still in high school Müller refused. Armstrong kept asking, and finally, in 1980, when Müller's son and daughter were both in college, Müller accepted. The director of the Zürich lab felt he could only spare Müller for three months. Müller did not think it would be worth his while to go for a visit of less than two years. Eventually a Helvetic compromise was reached: eighteen months, which in the end became twenty.

When Müller arrived in Yorktown in 1980 he was not yet an IBM Fellow; that honor would come in 1982. An IBM Fellow is one of a handful of distinguished scientists to whom the corporation gives a free ticket to follow their scientific fancy where it leads, with no regard for profitability. "Being awarded the Fellowship," Müller explains, "acknowledged that I was a guy who behaved like a Fellow already." Throughout his career he had always more or less done only experiments that interested him. If for some reason these experiments did not also interest IBM he would find a way to do them on the sly, using equipment purchased for sanctioned research. So, although he arrived at Yorktown Heights in 1980 not yet an IBM Fellow, he was treated like one by John Armstrong. "You can come in to the research center or you can stay home and study books, I don't care," Armstrong said. "Of course he knew that I would come every day," Müller recalls. Before long Müller was happily involved in several ongoing projects at Yorktown. One of these projects involved a large group of scientists and engineers at Yorktown who had, for several years, been working on what was known as the Josephson Computer Project, an effort—since abandoned—to build an extremely fast computer, about the size of a grapefruit, out of superconducting switches known as Josephson junctions. "Somehow I had developed a complex," Müller recalls, "because there was still this Josephson project going on and I was a solid state physicist and I didn't understand really what superconductivity was." So he started reading textbooks until two or three in the morning, and doing simple experiments, "like a young Ph.D." Within a short time Müller had mastered the literature of superconductivity. He performed a measurement on granular aluminum, a superconductor at very low temperatures, using a technique known as electron spin resonance that he had used in his studies of structural phase transitions. He wrote a paper, gave a talk at Harvard, and everyone agreed that it was a nice piece of work. Not yet earthshaking, but superconductivity had caught Müller's attention.

When he met Alex Müller for the first time, Johann Georg Bednorz was still a graduate student, associated with the Institute of Crystallography at the Swiss Federal Institute of Technology (ETH), Müller's alma mater. Bednorz was a quiet young man who was very good at growing crystals, stacking atoms in a regular array as easily as cannonballs. In the citadel of scientific rationalism the growing of crystals is still more art than science, a rare and essential knack.

Bednorz was born in the German city of Münster in 1950. From an early age he was interested in chemistry, and, like budding chemists everywhere, he had a basement laboratory in which he concocted solutions that changed color and powders that exploded. Bednorz thought about becoming a medical doctor, but his grades were not good enough to get him into one of the highly competitive German medical schools. Instead he decided to study chemistry.

Not long after transferring to chemistry Bednorz discovered that a lot of other students had had the same idea; classes were large and impersonal, not at all to his liking. After about a year of this he moved to the Institute of Crystallography to study solid state physics, where things were more *gemütlich*—warmer and friendlier. "It was kind of a family atmosphere there," he recalls. "There were a small number of students, with everybody helping the younger ones, and I felt quite at home from the beginning."

The study of crystals gave science its first entrée into the atomic world. Consider, for example, a crystal of ordinary salt. Large, cubic crystals of salt, the remains of ancient seas, occur commonly in nature. Cleave one of these crystals with a blade, or smash it on the ground, then look at the fragments. Each one, regardless of how small, is cubical, a reduced model of the original crystal. This was taken by the early atomists as evidence of the reality of atoms, which are stacked in regular patterns to make crystals. In the twentieth century solid state theorists reveled in the almost appalling regularity of crystals because this regularity simplified their equations, allowing them to solve problems involving trillions upon trillions of atoms with relative ease. For this reason most of what is known today about solid matter applies only to crystals. Amorphous materials, in which the atoms are haphazardly stacked, are only recently yielding to the mathematical attacks of solid state theorists.

Bednorz enjoyed growing exotic crystals, the delicate balancing of temperature and composition that nature could not match. He was soon growing crystals that no one else in Europe could duplicate, and he became fascinated with characterizing their physical properties—the distance between atoms, the way they transmitted light and heat, and their electrical resistance. While still a graduate student he began to work with scientists over at IBM.

Meanwhile, stealing time from his work on the scanning tunneling microscope, Gerd Binnig was doing experiments on superconductivity. He was particularly interested in strontium titanate, a peculiar superconductor that had been discovered in 1966. Strontium titanate is an oxide, a material containing oxygen, and at that time few oxides were known to be superconductors. As a practical superconductor strontium titanate was hopeless. Its critical temperature, the temperature at which it becomes a superconductor, was only 0.3 K, less than a degree above absolute zero.

The surprising thing was that strontium titanate conducted electricity at all, let alone that it was a superconductor. In order for a material to conduct electricity it must have a sea of electrons that are free to wander throughout the crystal lattice. These electrons come from the outermost shell of the constituent atoms of the crystals. Some atoms freely donate their outer electrons, and they are said to have positive valence; others have negative valences and are electron hungry. When an atom with a positive valence meets one with a negative valence, they fulfill each other's complementary needs and become chemically bound together. Oxygen is among the hungriest of atoms, and oxygen in a compound usually consumes all the free electrons in its vicinity, forming strong chemical bonds with its neighbors. After these bonds have formed, few free electrons are left over to wander the lattice, so oxides are most often insulators. But strontium titanate was a superconductor, which fascinated physicists. It was an oddball. It had 100 times fewer charge carriers, or free electrons, than other known superconductors. According to the then-accepted theory of superconductivity, the BCS theory, put forth by John Bardeen, Leon Cooper, and Robert Schrieffer in 1957 and justly counted as one of the crowning achievements of twentieth-century physics, the critical temperature is a very sensitive function of the density of charge carriers. So even a tiny increase in carrier density should send the critical temperature shooting up. Binnig thought he knew how to increase this carrier density, but to carry it out he needed somebody skilled in crystal growing, somebody like Bednorz, to whom he turned.

The idea was to dope strontium titanate with tiny amounts of the element niobium (to a chemist "doping" means adding a trace amount of a foreign element to the crucible in which the crystals grow). Niobium is similar in size to strontium, close enough that it would fit into the strontium sites in the strontium titanate crystal lattice. But niobium has more outer, valence electrons to contribute to the carrier sea. By doping with niobium the physicists hoped to increase the density of carrier electrons, which in turn should cause the critical temperature to soar. And this worked, up to a point.

When Bednorz grew crystals that were slightly doped with niobium the critical temperature did indeed rise. Increasing the level of doping also further increased the carrier concentration and, just as theory had predicted, the critical temperature increased further. Bednorz and Binnig continued adding niobium, until the critical temperature shot up to a sweltering 0.7 K. At this point, additional niobium destroyed any traces of superconductivity. It was the kind of thing that happens often in science, and especially in the science of superconductivity: Nature tempts and tantalizes—and then takes the back door. "This left Georg frustrated," recalled Müller, who at the time was Binnig's manager at IBM. "They had both had high hopes which disappeared." After writing up the experiment Binnig decided to abandon superconductivity to work full time with Rohrer on the new kind of microscope that would win them a Nobel Prize. "I always watched Binnig recording strange lines on an x-y recorder," Bednorz recalls, a little wistfully. "And I didn't understand at that time why he left the field of superconductivity and I was kind of disappointed."

Müller, meanwhile, had learned two useful things during his year and a half at the Thomas J. Watson Research Center. First of all, of course, he learned about superconductivity. More particularly, in his work on superconductivity he had concentrated on granular aluminum, a material that was at the center of a theoretical debate. Aluminum had long been known to be a superconductor; it has a transition temperature of 1.1 K. Granular aluminum, which consists of tiny grains of aluminum pasted together with aluminum oxide, is also a superconductor, but its transition temperature is much higher, more than 2.2 K. "Aha," Müller said, "the presence of the oxide must have something to do with it."

Müller began to believe that oxides were the way to go for "high-temperature" superconductivity. He talked to theorists about his idea, who patiently explained the reasons the critical temperature for any superconductor would probably never increase very much beyond the current record of 23 degrees K. The BCS theory of superconductivity could be used to prove that the absolute limit on critical temperature was about 30 K, or 40 K, at the most. There just was not room for much improvement, they explained. "That was more or less the last time I talked to a theoretician," Müller said. "Theory is good," Jean-Martin Charcot once remarked to Sigmund Freud, "but it does not prevent things from existing."

The second important thing Müller learned during his twenty months at Yorktown Heights was ceramics. IBM was having difficulties with the manufacture of multilayered ceramics, or MLCs, upon which microelectronic chips are mounted. After they scaled up their pilot MLC production line they were alarmed to find that some of the MLCs shrank more than others and were therefore useless. They enlisted Müller to help solve this problem, which he quickly did. He wrote a paper about his new method to control the shrinking that appeared in the *Journal of the Ceramic Society,* his only published work on ceramics.

By the time he arrived back in Zürich, in 1982, Müller

had reached some general conclusions about what kind of material could possibly become a high-temperature superconductor. First of all, he was convinced, it had to be an oxide, one containing either nickel or copper. He also had a notion about what crystal structure would be best. His intuition told him to look for materials in which something called the Jahn-Teller effect, a kind of distortion of the crystal lattice, was strong. With the benefit of hindsight, the Jahn-Teller effect appears to have been a kind of red herring that probably did not have a great deal to do with superconductivity in the new materials. In the end Müller proved to be right for the wrong reasons—which doesn't mean he was lucky. As Müller's hero Jung wrote, "The judgment of the intellect is, at best, only the half of truth." The Jahn-Teller effect, or better still, Müller's intuitive concept of the Jahn-Teller effect, was the star by which he steered himself through the infinite sea of chemical possibilities. The Jahn-Teller effect was, in a sense, the name Müller's intellect attached to his intuition.

By the summer of 1983 the general area to search was clear to Müller, but it was also clear that he would need someone to help him test the hundreds of possible compounds. He thought immediately of Bednorz, who was an experienced and talented crystal grower. Bednorz had recently received his doctorate after writing a thesis on a problem Müller had given him, and was now working full time at the IBM lab in Rüschlikon, growing crystals for himself and others. One day Müller dropped by his lab and told Bednorz about his idea of searching for new, high-temperature superconductors.

The two physicists met for about an hour, with Müller

doing most of the talking, explaining his ideas about the Jahn-Teller effect, the importance of nickel and copper oxides, and more. Müller came prepared to do a lot of convincing, but Bednorz stopped him with the word he least expected to hear. "Bednorz said in that same hour, 'Yes, I'm going to do it.' Which is exceptional. Because if you come to somebody who is younger with such a concept . . . well, it's a risky thing, with a lot of work ahead and no guarantee," Müller says. Even now he is a little dumbfounded, like a supersalesman with a product that sells itself. "Immediately he caught fire," he says, wonderingly.

As far as Bednorz was concerned, the arguments Müller had marshaled were quite strong enough, especially in light of the frustration he and Binnig had experienced in trying to elevate the critical temperature of strontium titanate. Both Bednorz and Müller realized that the search was far too speculative to carry on full time. Müller's instincts told him that it would probably be a good idea if they didn't let too many people know exactly what they were doing. Although by then Müller was an IBM Fellow with, in principle, an almost infinite freedom to follow his fancy where it led, he realized that most outsiders would consider a search for high-temperature superconductors to be quixotic, at best. So they continued on with their other projects, spending only about one-third of the time on their search, and while the search wasn't exactly a secret, they didn't go out of their way to advertise it either. Bednorz explained their secrecy afterward in a television interview: "We could afford to try ideas somebody else would have called crazy, and we could afford to follow these ideas, because we didn't tell anybody about it. So that's the reason why both of us, Alex and me, we were sitting together, we were discussing together, and we were cooking everything on a very small flame. So nobody realized what was going on. Because there was a certain risk—the project could be successful or it could be a disaster."

For the first two years they tried nickel compounds, scores of them. They scoured the literature of physics and chemistry for every nickel oxide compound that fit the criteria of their search. When they found one, Bednorz would make a series of samples, varying the proportions of the ingredients. The samples were usually pressed into tiny bars a few millimeters on a side. They painted four dots of silver onto the edges of the bars and to these dots they soldered wires to measure electrical resistance. Then they attached the sample to a probe equipped with an electronic thermometer and a tiny electric heater. The probe was at the end of a long copper arm that slid like a dipstick into a big silver Dewar of liquid helium. The wires from the probe were connected to a rack of electronics— power supplies, amplifiers, and, inevitably, an IBM PC, which controlled the experiment. When the sample was dipped into the liquid helium its temperature plummeted, and the computer produced a graph of resistance versus temperature. For two years the graphs were all more or less the same: They dipped and climbed and wiggled about, but none ever showed electrical resistance dropping to zero. None came even close.

"You probably can imagine what it was like, sitting in front of this bloody equipment, watching all the time the resistance going up and up and up," Bednorz said. After two fruitless years they decided to move from nickel compounds to copper. Still the resistance failed to drop.

After about two and a half years they decided to stop a while to think. "If you are concentrating all the time, if you are working like mad, you don't have the time to analyze, to look at the details," Bednorz says. "So sometimes it is important that you really go and think about what you have done." They repeated some of their experiments on nickel. And Bednorz decided to have one more look at the literature.

It was late in the summer of 1985 when, in a recent issue of *Materials Research Bulletin,* Bednorz came across a paper by Bernard Raveau and Claude Michel, two French chemists from the Laboratoire de Cristallographie et Sciences des Matériaux, at the Université de Caen. The paper described a copper-containing material, a ceramic consisting of barium, lanthanum, copper, and oxygen that met all the requirements of their search. Bednorz could not have ordered a better material. He went immediately to his technician and the two mixed up some samples of the French compound.

The synthesis of the French compound was relatively easy; it took only a few days, but Bednorz was in no particular hurry. He would soon have to give an important presentation to IBM's director of research, so he put the samples away without measuring them and concentrated on his other work. The presentation went well, but Bednorz was exhausted and took the Christmas holidays off. Finally, when he returned to his laboratory after New Year's he was ready to test the samples.

The resistance of the first samples he measured dropped by only 50 percent, but given their abysmal track record even this was promising. Over the next two days Bednorz refined his synthesis technique, varying the time he left the compound in the oven. That did the trick, because now the resistance of the samples dropped to zero at 35 K, 12 degrees higher than the record, almost too high to be accounted for by the BCS theory. "I stopped breathing at that time," Bednorz says, "and started thinking, did I do something incorrect? I came in the next day, repeated the measurement, and got the same result. I couldn't find any fault."

Bednorz was convinced that their sample was actually superconducting, but Müller counseled caution. He was only too aware of the dozens of false alarms in the past, the claims of superconductivity at enormous temperatures that later "went away" when nobody was able to reproduce them. Reputations were ruined by such things. Bednorz and Müller spent the next few months checking their experiment. Zero resistance at 35 K was a strong indication of superconductivity, but was it conclusive? If it were *absolutely* zero, zero to all decimal points, then it was. But no laboratory equipment could measure resistance that accurately, and the scientists knew that many uninteresting effects could mimic what they were seeing. So they repeated the experiment over and over. The effect would not go away. Nevertheless, as they knew, nobody would say that they had proved their case. So Bednorz and Müller decided to test for the other hallmark of superconductivity, the Meissner effect.

The Meissner effect was discovered by two German physicists, Walther Meissner and R. Ochsenfeld. (It should be called the Meissner-Ochsenfeld effect, but posterity is often more concerned with brevity than justice, so Meissner effect it is.) Before the discovery of the Meissner effect, scientists had assumed that all the peculiarities of superconductors were a direct consequence of the absence of electrical resistance. This seemed so reasonable an assumption that until the work of Meissner and Ochsenfeld nobody bothered to test it. In particular, nobody paid much attention to the magnetic properties of superconductors.

Magnetic fields, which may be visualized as looping lines of force that connect the poles of magnets, penetrate most metals quite easily. But, as Meissner knew, a magnetic field will not enter a metal that is already superconducting, because the magnetic field of the approaching magnet sets up an electrical current in the superconductor that flows without loss. This flowing current generates a magnetic field which precisely cancels the magnetic field within the superconductor. All this follows from the classical laws of electricity and magnetism applied to objects with zero electrical resistance. These same laws predict that any magnetic field penetrating a normal metal would become locked there when the material lost its electrical resistance. It seemed so obvious that nobody, until Meissner and Ochsenfeld, had bothered to do the relatively simple experiment. It therefore came as a total surprise when the German physicists found that under *no* circumstances could a magnetic field exist inside a superconductor; a magnetic field that penetrated a normally conducting metal would be thrust out if the metal should become a superconductor. In other words, a superconductor is a perfect magnetic shield. So to be absolutely certain that their new ceramic was a superconductor Bednorz and Müller had to verify the presence of the Meissner effect.

The device used to measure the Meissner effect is known as a magnetic susceptometer, and it can be readily purchased from companies that make scientific instrumentation. Any well-equipped superconductivity lab has one, but at that time IBM Zürich was not officially engaged in superconductivity research, so it was not so equipped. Müller and Bednorz immediately ordered one, but it would be months before it arrived. In the meantime they could have sent a sample to the Yorktown lab, which had the necessary equipment. "But we wanted to do this by ourselves," Bednorz says. "To have made the discovery and performed the final proof."

After two months of checking whatever could be checked without a magnetic susceptometer Bednorz and Müller decided it would be prudent to establish their priority by writing a paper. Others, they knew, had been working on oxide superconductors, and it was possible that they would be scooped. The trick was to publish a paper that would go unnoticed for a while, giving them time to do the susceptibility measurements. They chose, therefore, a journal that they felt was read by few superconductivity researchers, a German journal called *Zeitschrift für Physik* (*Journal of Physics*). They submitted their paper on April 17, 1986, giving it a title deliberately designed not to call attention to itself: "Possible High T_c Superconductivity in the Ba-La-Cu-O System." The journal would not arrive on most physicists' desks until about September, and Bednorz and Müller did not, as is customary, distribute preprints, not even to their colleagues at the other IBM labs. If they were right, the rest of the world would know soon enough. If they were wrong, the paper would sink with barely a trace.

In the following months, while waiting for the susceptometer to arrive, Bednorz and Müller gave a few small
talks to physics departments in Germany, trying their material out on the road, as it were. Their reception was
lukewarm at best. To their audiences, without measurements of the Meissner effect Müller and Bednorz were just
two more experimentalists who had almost certainly run
afoul of the subtle shoals of resistance measurement.

The susceptometer arrived in August. Bednorz and
Müller used this new piece of apparatus cautiously, resist-
ing the temptation to immediately measure a sample of
their new superconductor. First it had to be set up and
checked out. Then, to be certain they understood how to
operate it, they measured well-known superconductors.
Only then did they move on to the new material. They
found, as they suspected, that a large proportion of the
sample was superconducting. Mixed in with the superconducting phase there was also a normal phase in which the
atoms of lanthanum, barium, copper, and oxygen had
stacked differently and in different proportions. The non-
superconducting phase formed because they had not
started with exactly the right mix of chemicals to produce
a pure superconducting phase. The next step would be to
find the pure phase and determine its detailed atomic
structure, because that was where the secret of these new
superconductors was contained. From then on things
would move very quickly; this became a task that others
would complete.

Bednorz and Müller quickly wrote up another paper
describing their measurement of the Meissner effect. At an
industrial lab such as IBM, to make sure that proprietary
secrets are not divulged, papers must be cleared by re-

search managers before they can be published. Müller was an IBM Fellow and in effect his own research manager, but Bednorz needed the signature of his boss, Heinrich Rohrer. As it happened he found Rohrer to be in very good spirits. Twenty minutes before Bednorz approached him, on October 15, 1986, Rohrer had received a phone call from Stockholm informing him that he and Gerd Binnig would share the Nobel Prize in physics for their work on the scanning tunneling microscope. Jung might have called it synchronicity.

The self-taught Scottish physicist Sir James Dewar was one of the most ingenious experimentalists who ever lived. Among much else he invented the double-walled thermos, making possible the study of physics at extremely low temperatures.

ILLUSTRATED LONDON NEWS

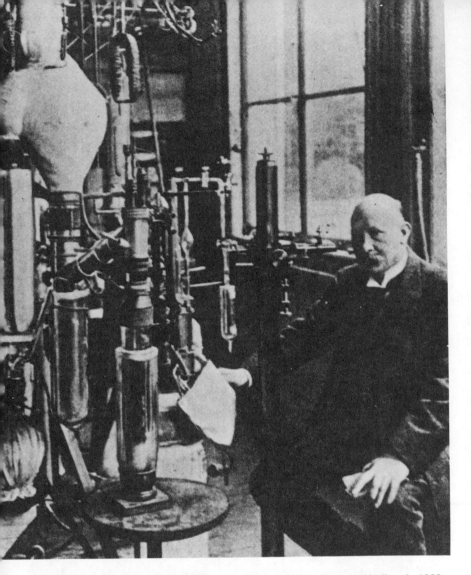

Kamerlingh Onnes and the apparatus with which he liquefied helium in 1908.

(above right) Heike Kamerlingh Onnes, "the gentleman of absolute zero," built the prototype of the modern scientific laboratory at the University of Leiden. His explorations of the properties of matter chilled to the temperature of outer space led him to the discovery of superconductivity in 1911, for which he would receive the Nobel Prize two years later.

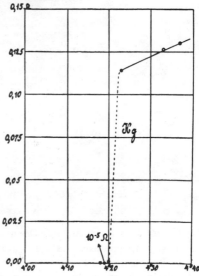

The discovery of superconductivity. The original graph made by Kamerlingh Onnes of the resistance, in ohms, of mercury (chemical symbol Hg) versus temperature. At just above 4.2 K the slowly falling resistance plunges suddenly to zero.

Superconductivity pioneer Bernd Matthias and his many collaborators discovered thousands of low-temperature superconductors. Just before his death in 1980 he told a colleague that oxide superconductors embodied "a completely new type of superconductivity."

AIP NIELS BOHR LIBRARY/PHYSICS TODAY COLLECTION

In 1957 John Bardeen, Leon Cooper, and Robert Schrieffer proposed a theory of superconductivity that had eluded physicists for over half a century. For this achievement they received the 1971 Nobel Prize in physics. The BCS theory cannot explain the new high-temperature superconductors, and theorists are racing to find one that does.

AIP NIELS BOHR LIBRARY

J. Georg Bednorz and K. Alex Müller in their laboratory at IBM's Zürich Research Laboratory—the outsiders who ignited the worldwide superconductivity fever.

COURTESY OF IBM

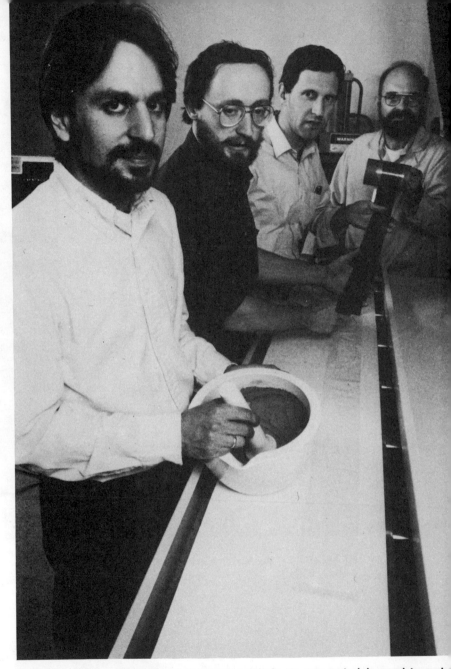

Robert Cava of AT&T Bell Laboratories grinds up raw materials used to make superconductor in a mortar and pestle. Bruce van Dover, Bertram Batlogg, and Da Johnson unfurl a roll of superconducting tape that, if perfected, might someday used to make magnets or cables.

AT&T BELL LABORATORIES

"Houston's newest superstar scientist," Paul Chu, became a media hero after the discovery of liquid-nitrogen-temperature superconductors.

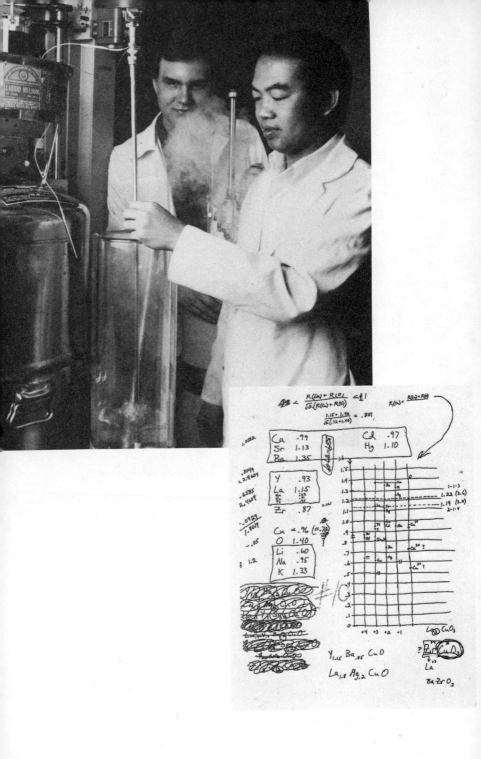

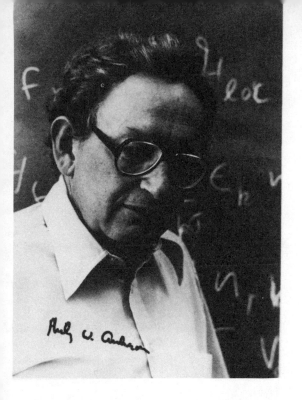

Nobel Prize–winning theorist Philip W. Anderson has proposed the leading theory to explain the new superconductors. Anderson's theory, which has not been experimentally confirmed, predicts that room-temperature superconductors should be possible.

AIP MEGGERS GALLERY OF NOBEL LAUREATES

(left above) **Maw-Kuen Wu and his graduate student, James Ashburn, at the University of Alabama in Huntsville. On January 28, 1987, Ashburn and Wu baked yttrium, barium, copper, and oxygen into a pale-green sample that became superconducting at well above the temperature of liquid nitrogen. "The pivotal contribution of Wu's team has been glossed over or ignored completely"** [*Science,* August 5, 1988].

UNIVERSITY OF ALABAMA IN HUNTSVILLE

(left below) **On the basis of this calculation, which was scribbled on the back of a homework assignment on January 17, 1987, Jim Ashburn decided to investigate compounds of yttrium, barium, copper, and oxygen. The first samples he mixed showed signs of superconductivity above the temperature of liquid nitrogen.**

JAMES ASHBURN

The "Woodstock of physics." On March 18, 1987, thousands of physicists crammed a ballroom at the New York Hilton to celebrate the coming of the age of superconductivity.

(right) Alex Müller, Paul Chu, and Shoji Tanaka, answering questions at the "Woodstock" meeting. Tanaka and Koichi Kitazawa were the first to confirm Bednorz and Müller's discovery, launching a worldwide race to find still better superconductors.

In the summer of 1988 Palmer Peters of the Space Science Laboratory in Huntsville, Alabama, discovered that superconductors to which silver oxide is added will not only levitate *above* magnets, but will hang suspended below (or to either side of) a magnet. The suspension effect results from magnetic lines of force that penetrate into the superconductor and become pinned there, tethering the superconductor like a child's balloon on an invisible string.

DENNIS KEIM, NASA

(left above) A magnet floats above a wafer of yttrium-based superconductor cooled in a bath of liquid nitrogen.

PHOTO COURTESY OF IBM

(left below) Easy as 1-2-3. When Paul Chu heard that Heidi Grant, daughter of IBM physicist Paul Grant, had demonstrated levitation to her eighth-grade class, he invited her to give a repeat performance to the advisory board of the National Science Foundation in Washington. This time she followed the "Shake 'n' Bake" method to make enough superconducting wafers to pass out to all the board members.

PHOTO COURTESY OF PAUL GRANT, IBM

The physical properties of solids are directly related to the three-dimensional arrangement of their constituent atoms. This is a computer rendering of the atomic structure of a ceramic superconductor. Physicists now believe that the secret of high-temperature superconductivity lies in the parallel planes of copper and oxygen atoms, which in this material are sandwiched in between layers of barium and yttrium atoms.

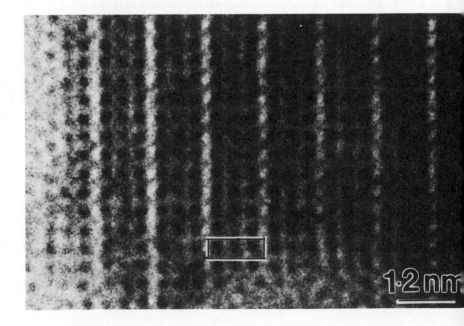

1·2 nm

(left below) In this electron micrograph of a ceramic superconductor magnified approximately 20 million times, individual atoms are visible. The vertical columns of lighter spots are copper and oxygen atoms, which are flanked by darker columns of barium and yttrium atoms. The box, which is only 1.2-billionths of a meter (500-millionths of an inch), encloses the three-layered unit that is repeated throughout the structure.

PHOTO COURTESY OF IBM

Electron micrograph of a single crystal of yttrium-barium-copper oxide, also known as 1-2-3, the first material to become superconducting at a temperature above that of the boiling point of liquid nitrogen.

PHOTO COURTESY OF IBM

The pattern of magnetic field lines surrounding a pair of powerful superconducting magnets is made visible by scattering iron nails on a piece of white
plywood.

GENERAL ELECTRIC RESEARCH LABORATORY

(right above) On July 22, 1987, President Ronald Reagan learned about
superconductivity from Edward Teller and other members of the White
House Science Council—Solomon Buchsbaum, executive vice-president
of Bell Labs, Ralph Gomory, senior vice-president and chief scientist at
IBM, Edward David, Jr., former vice-president of research at Exxon and
President Nixon's science adviser. "What has happened in the last
eight months the optimists thought would take two hundred years,"
Teller said, "while the pessimists were certain it would not happen at
all."

BILL FITZ-PATRICK, THE WHITE HOUSE

On July 28, 1987, at the Federal Conference on the Commercial Applications of Superconductivity, President Reagan watches a demonstration of a superconducting wire. Reagan gave the conference's keynote address during which he announced an eleven-point "Superconductivity Initiative." "Science tells us that the breakthroughs in superconductivity bring us to a threshold of a new age," Reagan said. "It is our task at this conference to herald in that new age with a rush . . . It's our business to discover ways to turn our dreams into history as quickly as possible."

MARY ANNE FACKELMAN-MINER, THE WHITE HOUSE

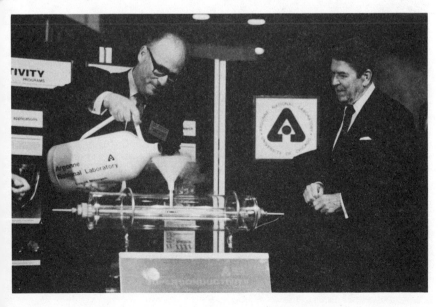

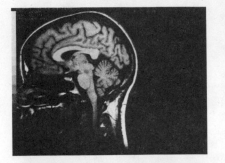

These images clearly demonstrate the advantages of MRI over other imaging techniques. Soft tissue, which is virtually invisible in X rays, can be clearly seen in MRI scans. Bones, which can cast an X-ray shadow, are accurately rendered in MRI scans.

IMAGES COURTESY OF GE MEDICAL SYSTEMS

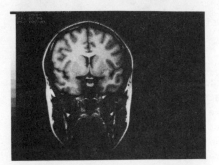

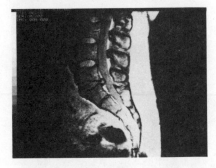

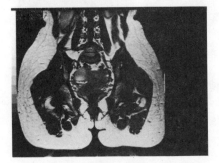

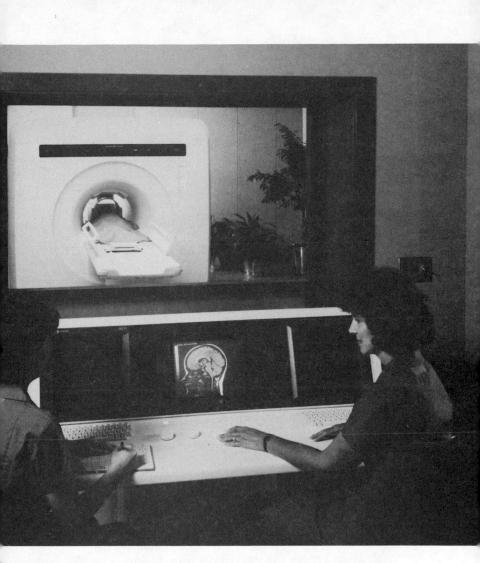

Magnetic resonance imaging, which produces three-dimensional images of the human body of unprecedented detail without radiation or injections of dyes, is the first commercial application of superconductivity. By 1987 over 1300 MRI systems, costing from $1.5 million to $2.5 million each, had been installed in hospitals around the world.

The superconducting rotor of General Electric's experimental electric generator is moved into place. The generator produced enough electricity to supply a community of 20,000 people. A conventional generator that produced this much energy would be at least twice as large.

GENERAL ELECTRIC RESEARCH AND DEVELOPMENT CENTER

Brian Josephson, the British physicist who discovered the Josephson effect, which forms the basis of the superconducting electronics used to make high-speed computers and sensitive detectors. After receiving a Nobel Prize, Josephson, age thirty-three, became a disciple of the Maharishi Mahesh Yogi and took up levitation—psychic, not superconducting.

PHOTO BY RAMSEY & MYSPRATT, COURTESY AIP NIELS BOHR LIBRARY

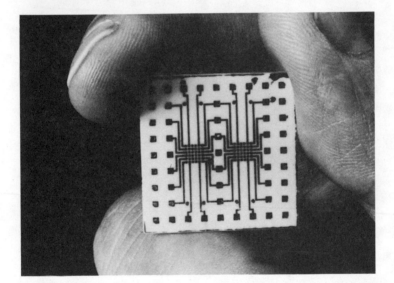

Because they have no resistance, superconductors can carry electrical impulses more rapidly than conventional wires. High-temperature superconducting wires like these that have been sprayed on a ceramic chip may someday be used to build faster computers.

PHOTO COURTESY OF IBM

(right above) The MLU-002 is a prototype of the Japanese magnetically levitated train. The onboard low-temperature superconducting magnets enable the train to fly 4 inches above the track. Speeds of up to 300 miles an hour are contemplated, which will reduce commuting time between Tokyo and Osaka to just one hour.

JAPANESE NATIONAL RAILROAD

Harold Weinstock of the Air Force Office of Scientific Research holds up a Japanese best-seller, a comic book about superconductivity. "Superconductivity is one of the major topics in the Japanese press. People appreciate it. They believe this is a big thing and they're going to win. And I'm not so sure I wouldn't bet with them if we continue the way we're going."

AMERICAN INSTITUTE OF PHYSICS

In January 1988 Zhengzhi Sheng and Allen Hermann of the University of Arkansas discovered a material containing thallium (an element previously used largely in rat poisons) that became superconducting at a record 125 Kelvin.

CHAPTER 5

SERENDIPITY

They sought it with thimbles, they sought it with care;
They pursued it with forks and hope . . .
—*The Hunting of the Snark*, Lewis Carroll

Bernd Matthias preached his gospel of empiricism to more than 150 students and collaborators, but he never had a more ardent disciple than Paul Chu. Chu, who was born in China in 1941, came under Matthias's spell as a graduate student at the University of California, San Diego, in the late 1960s. Ted Geballe, who often spent his summers in La Jolla, got to know Chu back then and remembers an eager, hardworking, and committed scientist. "He was always looking for the big discovery," Geballe recalls. "And not afraid to make a mistake. If there were two ways of interpreting something, he would take the optimistic way and see the consequences."

For Chu the "big discovery" had always meant just one thing: a high-temperature superconductor. Pushing the record critical temperature up by even a fraction of a degree would be gratifying, but Chu dreamed, as did many other scientists, of much higher temperatures. Theoretical investigations based on the BCS theory seemed to place a ceiling of about 30 K on the critical temperature, but, despite its many successes, the theory could still be wrong, or at least incomplete. Perhaps, as some theorists had conjectured, there were alternative, novel mechanisms of superconductivity. Matthias had taught Chu that it was not a good idea to take theorists too seriously, and Chu was a good student. Chu came to the University of Houston in 1979 and established a small group of superconductivity researchers consisting of about a dozen students and faculty. His laboratory was tiny and cramped; the only office some of his students had was a desk in the hallway. Chu's research was supported by grants from the National Science Foundation and the National Aeronautics and Space Administration, but his yearly funding never topped $100,000, which, even by the modest standards of superconductivity research, was a pittance. Once one of his graduate students accidentally opened a valve and in an instant released $1,500 worth of helium gas to the atmosphere. "We all make mistakes," Chu said. But even a small error could cost a significant fraction of his yearly budget.

Like writers and artists, the work of most scientists contains themes that recur throughout their careers. Chu's own leitmotif was pressure; he liked to squeeze superconductors to see how their properties varied as their atoms were pressed more closely together. The device Chu used to do the squeezing was called a diamond anvil, essentially a tiny vise with diamond jaws, which could, at the turn of a screw, re-create pressures found deep within the earth. Often the squeezing would raise the critical temperature, and then Chu would try to mimic this effect by replacing the atoms in the compound with other, smaller atoms, effectively shrinking the original lattice. Chu and his group managed to squeeze a lot of interesting physics out of superconducting crystals with the diamond anvil, but the big discovery always eluded him.

Immediately before coming to Houston Chu worked for a time at Cleveland State University with Alexander P. Rusakov, a visiting Soviet physicist. Rusakov and his colleagues in the Soviet Union had claimed that copper chloride, when compressed, became superconducting at a sultry 140 K. Their claim was almost universally doubted by scientists outside the Soviet Union, and especially by Chu's mentor, Matthias, who told the *New York Times* that the Soviet claim was "probably deliberately meant to deceive." Chu, working with Rusakov, squeezed a sample of copper chloride to 40,000 atmospheres and claimed to see a 7 percent superconducting effect. "If this is superconductivity, it must be something very unusual and complicated," Chu told the *Times.* Over the years Chu continued to work on copper chloride with Ted Geballe and others, but, superconducting or not, the material did prove to be very complicated; it appeared to be highly unstable and repeated measurements gave different results. Marvin Cohen, a Berkeley theorist, derided this research, saying that it belonged in the "Journal of Irreproducible Results."

After coming to Houston Chu became intrigued by oxide superconductors, those oddities that did not seem to fit into the standard picture of superconductivity. In 1975 a scientist at Du Pont named Arthur Sleight discovered an intriguing oxide, a ceramic made from barium, lead, bismuth, and oxygen, that became superconducting at 13 K. In the next decade groups all over the world would try to improve on Sleight's compound, but none would succeed. Bernd Matthias was one of the staunchest believers in the possibilities of oxide superconductors. Among those with whom he shared his enthusiasm for oxides was Rustum Roy, a physical chemist at Pennsylvania State University. Roy spent five fruitless years trying to improve on Sleight's recipe. In October 1980 Roy visited Matthias and told him: "I'm not wasting any more time on these damn oxide superconductors."

"Don't give up, Rusty," Matthias told him. "This is a completely new type of superconductivity." Matthias would not live to see his intuition confirmed; that night, on October 27, 1980, after playing a game of tennis, he died in his sleep.

Paul Chu also believed in the promise of oxide superconductors, and, unlike Roy and others, he never gave up. He and his students constantly scanned the world literature for strange and unusual new compounds that might, upon cooling, become superconductors. He played long shots and hunches—almost nothing was too bizarre. One night in 1982 he dreamed that sodium sulfide would be a high-temperature superconductor. For the next two weeks he and his students mixed up dozens of new compounds, baked them and squeezed them and cooled them to the brink of absolute zero. None of the samples was a superconductor; none was even a conductor. In the summer of 1986, after years of such frustrations, Chu gave himself an ultimatum: He vowed to his wife that if he didn't find a high-temperature superconductor within the next three years he would give up the search. Chu did not then realize that the big discovery he had been working toward for his entire career had already been made. The issue of *Zeitschrift für Physik* containing Bednorz and Müller's article had already been printed. Within a few months Chu would arrive at his lab at dawn, as was his custom, and find a photocopy of their paper on his desk. A new superconductor, an oxide no less, at 30 K. How could he have missed it?

Later, after he had become, as an editorial in the *Houston Post* would put it, "Houston's newest superstar scientist," Chu would reenact the moment for the cameras of the PBS science series "NOVA." In this reenactment Chu does not hesitate; the skepticism with which others greeted the Zürich paper apparently did not trouble Chu. He immediately called his group together and made the grim announcement: "The Swiss got it."

Before long Chu and his students had duplicated the work of Müller and Bednorz, but they did not take it any further. Like the IBM physicists, Chu had found that the resistance of his sample became zero at about 30 Kelvin. But, as he well knew, resistance measurements are tricky: The difference between a zero resistance and a very tiny resistance can be difficult to discern, and only samples with absolutely no resistance are superconducting. To demonstrate convincingly the existence of superconductivity requires observation of the Meissner effect. Bednorz and Müller had not had the equipment necessary to perform this vital measurement, which is why they chose to publish their paper with its equivocal title, "Possible High T_c Superconductivity in the Ba-La-Cu-O System" in *Zeitschrift für Physik,* a relatively obscure journal. Chu apparently did not perform this measurement either, preferring instead to spend his time trying to improve on the Swiss recipe.

In December Chu tore himself away from his Houston lab long enough to attend the Materials Research Society meeting in Boston, where he met Koichi Kitazawa and learned that he was not the only one to have noticed Müller and Bednorz's paper. The Japanese had carried the Swiss work further than he had; they had measured the Meissner effect. Chu rushed back to his lab in Houston, convinced that he was now in a race.

It *was* a race, one of the most remarkable in the history of science. Within a few weeks dozens of laboratories around the world were frantically mixing up variants on the Zürich theme of lanthanum-barium-copper-oxide. The new superconductor was easy to make; it required few tools that would not have been found in an alchemist's laboratory 400 years ago. The raw ingredients are mixed and ground with a pestle in a mortar and baked in an oven for a few hours. The whole process may be repeated and some of the steps might be varied a little, but that is all there is to it. Anyone could do it and almost everyone did.

Before returning to Japan Kitazawa made a brief stop at Bell Labs to tell the scientists there who had not been to the MRS meeting about his lab's latest results. The Bell scientists, led by Robert Cava and Bertram Batlogg, started right in to search the uncharted reaches opened up by Bednorz and Müller's discovery. Like Chu, they turned at once to the periodic table to find elements similar to lanthanum, barium, and copper that might be substituted without too badly distorting the superconducting structure. The procedure they followed was straightforward and tedious because there were so many parameters to vary: the proportions of the elements, the time in the oven, the temperature, the rate of cooling. The scientists at Bell, like their counterparts in Houston, Tokyo, Zürich, Beijing, and elsewhere, began camping out in their laboratories, grinding and baking hundreds of samples, attaching them to probes, and plunging them in Dewars filled with liquid helium. As the temperature dropped, their eyes were fixed on the pen that drew a line on a slowly unscrolling graph of resistance, hoping for the wiggle that was the prelude to superconductivity, and hoping that they got there first.

At Bell Labs, Batlogg and Cava were working their way up the periodic table. One of their first ideas was to find substitutions for the barium in the Zürich superconductor. Strontium occupies the space in the periodic table directly above barium, so they tried making compounds of lanthanum, strontium, copper, and oxygen. And almost immediately they succeeded in finding a new superconductor with a critical temperature of around 42 K, a temperature that was extremely hard to reconcile with the BCS theory. It was becoming increasingly likely that another mechanism of superconductivity was at work in these new compounds, and if this were so nobody could say how high the critical temperature would go.

At the MRS meeting Chu had run into his former student, Maw-Kuen Wu, who had his own superconductivity group at the University of Alabama in Huntsville. The two decided to begin an informal collaboration, so they could share information and not duplicate too much work. Wu's group hit upon the strontium substitution at around the same time Bell Labs had, but the Bell scientists had hand-delivered their paper to *Physical Review Letters* and the two papers appeared back to back in the same issue. The Houston and Alabama scientists were disappointed by the dead heat with Bell, but the race, they suspected, was not yet over. Higher critical temperatures were probably possible.

To guide his search, Chu resorted to his favorite technique, and began applying pressure to samples of the lanthanum-based superconductor. The results were immediately encouraging; the critical temperature shot up. In successive experiments he saw the critical temperature rise to 40, 52, 57, and, for one tantalizing run, to over 70 Kelvin. For some reason that Chu did not understand, the critical temperature rose as the atoms drew more closely together. The problem now was to counterfeit the pressure effect chemically by substituting smaller atoms for the lanthanum components of the superconductor. A quick look at the periodic table convinced Chu that an element called yttrium, a member of the rare earth group, could replace lanthanum. Toward the end of January 1987 Chu and his students at Houston began churning out dozens of compounds made of yttrium, barium, copper, and oxygen.

At the same time, in Alabama, Wu and his two graduate students were working their own way across the periodic table. They tried substituting calcium for the barium, which lowered the critical temperature, and then magnesium, which did not work at all. On July 17, one of Wu's graduate students, twenty-three-year-old James Ashburn, scribbled a simple calculation on the back of a homework assignment that also suggested, for reasons that had nothing to do with Chu's pressure effect, that yttrium might work. Wu's lab did not stock yttrium at the time, so Ashburn had to order some from a chemical supply house, and the order took another week to arrive. On January 28 he mixed up a batch of yttrium, barium, copper, and oxygen and put it in the oven to bake overnight at 1000 degrees Celsius. The sample that was pulled from the oven was greenish, not a promising sign, since the laws of solid state physics unambiguously forbid any green conductors, let alone green superconductors. But mixed in with the green was a black material that they felt was worth investigating.

According to the time stamped on the computer printout the Alabama researchers did their first run on the greenish material on January 29, at 2:08 P.M. They placed the sample in a Dewar of liquid helium, and watched the electrical resistance, which was plotted on the computer screen, drop with falling temperature. When the temperature hit 89 K the graph started to bend over, and it continued to fall until it hit zero at around 50 K. Before allowing themselves to get too excited they checked the thermocouple to see if it was registering the temperature correctly, then tried other samples from the same batch. By early that evening Ashburn had mixed up some more of the material, and this time the results were even better. Trembling with excitement, Wu called Houston to report the good news. That night, driving home from the lab, Ashburn thought happily about all the wonderful applications of these new superconductors. The next day he and Wu flew to Houston to check their results on Chu's more sophisticated equipment. Over the next few days the physicists refined their recipe and began routinely to see superconducting transitions in the low 90s, well above the temperature of liquid nitrogen. The superconductivity revolution had arrived.

On January 30, Chu called Myron Strongin, the editor of

Physical Review Letters, with the news. "We've got it," he said. *Physical Review Letters,* like most scientific journals, is a refereed journal. This means that papers submitted for publication are first reviewed by anonymous "referees," scientists with the expertise to judge the quality of the submission. Chu was nervous about leaks, so he asked Strongin if he could submit the paper for refereeing without the formula, which would be inserted only on the day the issue was to be printed. Strongin told him that would be impossible; the referee had to be told the correct ingredients, or he would be unable to make a proper judgment. Chu reluctantly agreed, and promised Strongin that two short papers would soon follow.

Chu quickly threw together a paper for the group, and gave it to his secretary to type and send via Federal Express to the journal's offices on Long Island. *Physical Review Letters* received the paper on February 6, a Friday. The paper contained errors, possibly typographical, that would return to haunt Chu.

The range of expertise and equipment necessary in modern scientific research makes it essential that scientists, especially from small labs like Chu's, form networks of collaboration. When Chu needed to perform critical magnetic measurements he often turned to Chao-Yuan Huang, who at the end of January was just completing a move to Lockheed in Palo Alto, California. On Saturday, January 31, Huang returned to a friend's house, where he was temporarily living while settling into a place of his own, to find an urgent message to call Chu. "So I call back and Chu said they had discovered a new high-temperature superconductor. I said, 'What temperature?' He said, 'Above eighty, maybe higher.' I was really suspicious." Chu wanted Huang to do the experiment the next day, a Sunday, but, because of the move, Huang could not get to it until Thursday, February 5. When he did his suspicion vanished. He saw an unmistakable signal in the magnetization curve at 90 K. Huang became very excited, because this was more than a discovery, it was redemption. "Paul and I used to work on copper chloride," he said, "and we got a lot of blame and attack."

On Valentine's Day Huang and one of Chu's students, Pei-Herng Hor, flew to Boston to make additional measurements with the big magnet at the Francis Bitter National Magnet Laboratory, which is affiliated with MIT. Simon Foner, director of the lab, watched as Huang poured liquid nitrogen into the Dewar containing his sample. "Is that all you need?" Foner asked.

Word of Huang and Hor's remarkable doings spread rapidly, and soon a crowd of incredulous physicists gathered, all with the same question: "Is this thing for real?"

"Yes," Huang said, "providing that the liquid nitrogen you gave me is *real!*" The scientists laughed, sharing in the giddy rush of discovery, and then rushed to the phones to indulge themselves in the pleasure of spreading good news.

The next day, February 15, through the offices of the National Science Foundation, Chu officially announced that, in collaboration with Wu at the University of Alabama, he had found a superconductor with a critical temperature somewhere around 90 K. From then on Chu would be the spokesman for the collaboration and the role of the Alabama scientists would be mentioned less and less in articles about the breakthrough.

On January 12, before making any yttrium-containing

materials, Chu met with University of Houston lawyers to file a broad patent that included, among much else, the still-untried formula that in a little over two weeks would yield the 90-K superconductor. What protection, if any, this patent gave Chu will probably be the subject of interminable litigation. So, ever cautious, Chu chose not to divulge the formula for making the new superconductor at his press conference. For that the world would have to wait a couple of weeks, until March 2, the day the paper describing the discovery would appear in *Physical Review Letters.*

The day after the official announcement, news of the discovery was featured in many of the country's leading newspapers. Chu would soon become adroit and careful when handling the press, but on the day of his announcement he gushed to a *Houston Post* reporter, "It's going to be great. It's really amazing."

And it was. Everywhere scientists were desperately trying to guess the magic formula. The international grapevine crackled with rumors. One scientist got hold of a Chinese paper that he believed divulged the correct composition; he couldn't make sense of the ideograms, but the symbols for the chemical elements are universally written in Roman characters. One that kept recurring was Pb, the symbol for lead. So a lab full of scientists sacrificed a night's sleep to mix up lead compounds. They later found that the Chinese had been using lead, which is a well-known superconductor, for calibration.

Jean-Marie Tarascon, a solid state chemist at Bellcore, the research arm of the seven regional Bell operating companies in Red Bank, New Jersey, was almost driven mad by the constant barrage of rumors. Each had to be checked out, regardless of how absurd it sounded, if for no other reason than to satisfy the lab management. A picture of Chu holding a wafer of his new superconductor appeared in *Time* magazine, and in the article he was quoted as saying that it was green. It sounded unlikely to anyone familiar with the laws of solid state physics, but then, so had a 90-K superconductor. Nickel oxide, rumor had it, was the secret. To Tarascon this was insanity; nickel made no chemical sense at all, so he did not try it. Then one day he arrived at the lab to hear that Chu was about to hold a press conference announcing that the 90-K material contained nickel. "When I heard this I was completely depressed," Tarascon recalls. "It turned out that we were out of nickel oxide here, so we called every lab around to try to get some." Tarascon mixed up a batch of nickel compounds and convinced himself that his intuition had been correct. Nickel was not the secret ingredient. And the rumored press conference never took place.

One of the most troubling rumors was that the secret ingredient was ytterbium, a rare-earth element with the chemical symbol Yb. The origin of this rumor turned out to be easy to trace: It was found in the manuscript describing his discovery that Chu's secretary sent to *Physical Review Letters.* The contents of papers submitted to *Physical Review Letters* are, in theory, never divulged to anyone but the editor and the paper's referees. Chu had been worried that the journal's editors would choose one of his rivals as a referee, so he requested a referee of his own choosing. His choice was John Hulm, an old and trusted friend. Hulm agreed, and within a few days he read the paper and found it fit to print. This did not mean that the paper was correct, only that it was reasonable, free from glaring mistakes. As it turned out, it contained a few errors —or perhaps red herrings.

Chu told reporters that his secretary had accidentally typed Yb whenever he had written Y, the symbol for yttrium. In addition—she must have been having a very bad day—she had incorrectly typed the proportion of "ytterbium." Nowhere in the paper was the name of the element spelled out. On February 18, the last day that corrections could be accepted, Chu called *Physical Review Letters* and told them he had just spotted the typos. The version of his paper that was set in type was correct.

The offices of *Physical Review Letters* proved to be as full of leaks as the Pentagon. Before long the word was out that Chu's secret substitution was ytterbium. Ironically, ytterbium *can* be substituted for yttrium and still give a 90-K superconductor, but at the time not even Chu knew this. Though many scientists have admitted to having heard the ytterbium rumor, none say they actually followed up on it before the publication of Chu's paper. At any rate, it is certain that nobody was able to get ytterbium to work.

Chu insists that the error in his manuscript was nothing more than a typographical error perpetrated by his secretary, but his story is disputed by Wu's student Jim Ashburn. Ashburn recalls a meeting at the University of Houston that was attended by Chu, Wu, and a few of Chu's graduate students to decide upon a plausible "typo" to put in their *Physical Review Letters* article. It was an amusing game, and the scientists laughed a lot. Someone suggested beryllium, but this idea was vetoed because that element is highly toxic; they did not want to hurt anyone, just throw them off the track. And, although the scientific ethics are questionable, the ruse worked. "It's strange," Chu slyly told James Gleick of the *New York Times,* "yttrium and ytterbium, they both start with 'Y.' People kept on calling me and saying 'Ytterbium's not working.' I just said 'Oh? It's not working?'"

Ironically, Roy Weinstein, dean of the College of Natural Sciences and Mathematics at the University of Houston, let slip the correct formula to a *Houston Chronicle* reporter, who included it in an article on February 16. Apparently, no researchers elsewhere saw the *Chronicle* story.

Chu's sometime colleague Robert Hazen, a scientist at the Carnegie Institution, of Washington's Geophysical Laboratory, told *Superconductor Week,* one of the many newsletters that would spring up in the wake of the Houston-Alabama discovery, that "although there's no hard evidence, we're sure . . . another 'spy'—a Chinese science attaché . . . had been following Chu around in press conferences that week."

One way or another, by late February a group of Chinese scientists at the Institute of Physics in Beijing had managed to make some 90-K superconductors containing yttrium. Their triumph was reported on the left corner of the front page of the *People's Daily.* The next day Jonathan Sun, a Chinese graduate student working in Ted Geballe's group at Stanford, read the article, which divulged the formula of the new, high-temperature superconductor, *kow whon tsan tao* in Chinese. By four o'clock that afternoon Sun had mixed up a batch using the *People's Daily* recipe, and popped it in the oven. (Researchers would later discover that these new superconductors, like TV dinners, could be made rapidly in microwave ovens. Sun's mix had to bake overnight.) The next morning, sure enough, the sample was superconducting at 90 K. "In this field," Sun admits, "being lucky is better than being smart."

Sometimes, in fact, to be smart is to be unlucky—at least when it comes to superconductors. Ever since Geballe had told him about the excitement at the MRS meeting, Jean-Marie Tarascon and his group at Bellcore had been mixing and baking hundreds of compounds. The first thing he would do after pulling them from the oven was to determine if the sample consisted of one or many phases. Tarascon was a chemist, not a physicist, so he rejected multiphased samples. "Generally, when you are a solid state chemist, when you see a multiphase compound you don't waste your time to look at the physical properties. Because it's garbage," he explains. Worse, many of his multiphased samples were green and so could not be metals. Garbage or not, Tarascon did not throw these samples away. He put them in a drawer on the off chance that someday they would be useful. Toward the end of February Geballe, a consultant to Bellcore, was visiting Tarascon's group in their New Jersey labs. On the 26th he received a phone call from Jonathan Sun, who had just finished reading the *People's Daily*. The answer is yttrium, Sun told Geballe, who passed the news on to Tarascon. Tarascon remembered the green samples that had been sitting in his desk drawer for over a month. Tarascon quickly pulled one out and tested it. At 90 K, as advertised, its resistance vanished. The Bellcore scientists hastily wrote a paper and drove it to the *Physical Review Letters* editorial offices at Brookhaven National Laboratories on Long Island, but it arrived well after Chu's had been received, too late to become anything more than a footnote in the history of science.

By then, other groups, aided by rumor and intuition, had also guessed Chu's closely guarded secret. The certainty that a 90-K superconductor was possible made the search easier, if only psychologically.

By the end of February a kind of frenzy had seized scores of the world's leading scientists. Chu's group might have reached 90 Kelvin first, but it seemed almost certain that this was but another step on the road, that the competition would continue and it would be fierce. The compounds were so simple to make that virtually anybody could do it, so the universe of competitors had expanded to include scientists who had never worked in the field of superconductivity before. Bell Lab's Robert Cava told "NOVA": "When this thing came, for me it was my chance to really make an impact on things. It's an opportunity that maybe comes around once in a scientist's career—or it never comes around in your career. And to be in the right place at the right time made me really want to go for it."

One of those who now decided that he was in the right place at the right time was Paul Grant, a senior scientist at the IBM Almaden Research Center in San Jose. Grant had heard rumors of Bednorz and Müller's discovery on the IBM grapevine, but, without independent confirmation, he had not taken them seriously. In December, a week after the Boston Materials Research Society meeting, word of the talks by Kitazawa and Chu reached him in San Jose. Grant's curiosity was finally aroused; he sensed that this thing, which he had all but ignored for months, was going to be very big, and so he leaped into action.

Paul Grant is a flamboyant figure, especially among scientists. He is tall and thin, with a mass of prematurely white hair and a flowing beard that makes him look like a cross between Buffalo Bill and Merlin. One day, after superconductivity had become a media event, he was standing in a supermarket checkout line, waiting to buy about a dozen issues of *Time* and *Newsweek*. A little embarrassed by this purchase, he felt obliged to explain to the woman ahead of him that for the first time in his life he appeared in both these magazines. The woman sized him up: that beard, the hair, those wild eyes, that deep, impressive voice. "You're one of those TV evangelists, aren't you," she said. Grant would frequently tell this story from the podium at conferences, and it would get a good laugh. Then he would say, "I guess I am an evangelist in a way. For superconductivity." When Grant was finally convinced of the reality of the Zürich experiment he did not, as others might, reach for the phone. Instead, he checked his laboratory budget, and, seeing there was enough surplus, he purchased a round-trip ticket for Zürich. When he arrived he found that Müller had gone off skiing, but Bednorz was there, "working like crazy. I spent pretty much the whole day with him, and was blown away by how much data they had," Grant recalls. "That's when I became a true believer."

In February, Grant, like everyone else, was trying desperately to second-guess Chu, but with no success. Toward the end of the month, on Thursday the 26th, Chu gave a seminar at the University of California, Santa Barbara, where he promised to share some of his data—not the formula, just the graphs—for the first time. Chu's audience consisted of a very select group of some of the foremost names in superconductivity, including two Nobel Prize winners. "Paul got up and showed a foil of the resistance versus temperature," Grant recalls. "I was just floored." Everyone was; Chu's data left no doubt. Chu showed some X-ray data that indicated his sample was a jumble of at least two phases; under a microscope the matte green proved to consist of a mix of tiny green and black crystals. Grant was worried that the superconductivity took place at the interface between the two types of crystals, and would therefore be complicated. Grant was the only experimentalist in the room, the only one who had a reason to be discouraged by Chu's remarks. The theorists included Robert Schrieffer, one of the authors of the BCS theory, and Philip W. Anderson, a brilliant theorist who had received a Nobel Prize in 1977. Anderson had already come up with a theory of how these new superconductors worked and he tried to reconcile it with this latest discovery. Even to his fellow theorists, Anderson's theory, which he called the resonating valence bond theory, was difficult to grasp, and a lot of that seminar was devoted to arguing about it. To the theorists Chu's announcement was terrific news: BCS theory, it appeared, was probably wrong, and whoever set it right would almost certainly win a Nobel Prize. To Grant, the experimentalist, Chu's news meant an opportunity missed, and he was determined not to miss any more.

That night at dinner the scientists were in superlative spirits. They joked about having a meeting in Aspen the following winter where they would schuss down the slopes levitated on superconducting skis; by then someone would surely have discovered a superconductor that operated at the temperature of frozen water. Chu told them that he had instructed his secretary to send ten preprints of his paper via Federal Express to selected scientists— Grant was not on the list, and was disappointed. But Ted

Geballe at Stanford—just down the road from IBM Almaden—was, and Grant began calculating how he could get a glimpse of Geballe's copy.

The next morning Grant called Geballe and asked him to let him know the minute he received the preprint. Geballe called after lunch and read him the figures, so much yttrium, so much barium, so much copper. A few minutes later an old colleague of Grant's, Richard Greene, called from the IBM lab in Yorktown Heights, New York, with the same numbers.

That night Grant and his colleague Ed Engler snatched their first samples hot from the 900 degree Celsius oven. They wore gloves and tossed the samples from hand to hand, like a hot potato, to cool them off rapidly. Then they measured them and found that they were not superconducting. The scientists worked all the next day, but the resistance curves only barely began to dip at 90 K. It was maddening. The following morning, a Sunday, at eight o'clock, Grant's phone rang at home. "Are you coming into the lab today?" Mike Ramirez, a laboratory technician, asked. "I've got some samples here with zero resistance at ninety-two degrees," Ramirez said. Grant got in his car and quickly drove up the winding road to the lab in the Almaden hills.

Grant wondered why this run had worked when the others had not. Then it hit him: He and Engler had been too anxious. In order to be superconducting the samples had to be cooled slowly. The trick was to shut the oven off and let it cool overnight; oxygen entered or left the lattice depending on the rate of cooling and, for reasons still not understood, the precise oxygen content is critical to superconductivity. Grant called up Greene in Yorktown in the middle of a run, and narrated it like a sportscaster: "It's at a hundred degrees... it's beginning to drop now ... Greene, it's going to zero at ninety-two degrees!" The two scientists could not contain themselves any longer. They began to laugh, hysterically, uncontrollably. "It was just so mind-blowing," Grant says.

By the end of the day Grant and Engler had found that they could substitute the rare-earth element lutetium for yttrium and not alter the transition temperature appreciably. Their samples, like Chu's, were a matte green mixture of black and green particles. The results were so consistent that Grant became convinced that they had nothing to do with the interfaces between the two particles—because these varied in form from one sample to the next—and therefore must be confined to one phase. And unless the last fifty years of solid state physics had been badly mistaken, there was no way the superconductor could be the green phase. The problem, then, was to find the composition and structure of the black phase. Finding the structure was a fairly straightforward, almost mechanical, task for those who had the proper equipment. It was a race that would be won not necessarily by the smartest, but the quickest and the luckiest. But that did not matter to the contestants: First was first, and science has few rewards beyond priority. The first group to find the structure of Chu's new superconductor would be rewarded by nothing more than a footnote in future papers, but at least they would get that footnote. Grant phoned Myron Strongin at *Physical Review Letters* on Monday, March 2, to tell him that they had confirmed that Chu's material was a superconductor, as claimed. Strongin told him that another group had called Sunday. (It was Cava's group at AT&T

Bell Laboratories, but Strongin did not tell Grant that.)
Later that day a young Ph.D. candidate at Almaden named
Robbie Beyers began the X-ray diffraction studies necessary to determine the structure of the phases contained in
the lab's superconducting sample. Beyers worked quickly,
and within a day he had determined the structure and the
composition of the green and the black phases. The green
phase contained its yttrium, barium, and copper in the
ratio of 2-2-1. In the black phase the ratio was 1-2-3. Start-
ing with these compositions, Engler and Grant were able
to produce a pure green sample, which was an insulator,
and a pure black superconductor. They couldn't help
laughing. Making this superconductor really was easy: as
easy as 1-2-3.

Even before Beyers had completed his analysis Paul
Grant was writing the paper detailing the structure of the
new superconductor. When Beyers was through, Grant
filled in the last numbers and then hopped in a jet to hand-
deliver it to *Physical Review Letters.* But he was too late
by hours. Bell Labs had gotten their analysis to the journal's editors first, and after them came Grant's colleagues
at Yorktown Heights.

It does not ultimately matter which group discovered
the structure of the new superconductor; it was a straightforward problem that required no particular creativity,
brilliance, or insight to solve. Similarly, after the breakthrough by Bednorz and Müller the discovery of the liquid-nitrogen-temperature superconductors was probably
inevitable. Nevertheless, priority and the tiny place in history that it bestows are among the few public rewards of
a scientific career. Maw-Kuen Wu and James Ashburn are

properly listed as the first authors of the paper in which the discovery of the 90-K materials was announced. But after this something strange started to happen. The media began to ascribe the discovery solely to Chu and the University of Houston; the crucial role of the University of Alabama was mentioned less and less frequently. When they were mentioned at all it was as if they were little more than technicians operating under Chu's orders. Robert Hazen, in his book *The Breakthrough,* states that Wu's team was "acting as an extension of the Houston lab." In "NOVA" 's reenactment of the discovery the crucial contributions of Wu and Ashburn are not mentioned at all. For months Wu deliberately kept a low profile out of respect for his former teacher. As a graduate student Ashburn felt powerless to set matters straight. He began to refer to the discovery as "the UAH" (University of Alabama in Huntsville) disaster, an association that was strengthened in his mind by the coincidence of its having occurred one year to the day after the space shuttle *Challenger* exploded. He, too, felt as if he had been blown out of the sky.

At Penn State Rustum Roy was so upset by the way the media were bestowing credit for the discovery that he began to write letters to *Science* magazine, none of which was ever printed. But the editors of *Science* decided to investigate his charges. An article appeared in the August 5, 1988, issue that showed that "the pivotal contribution of Wu's team has been glossed over or ignored completely."

CHAPTER 6

GREAT EXPECTATIONS

Trains that will seem to run faster than a speeding bullet, magnets more powerful than a locomotive, home computers able to do your income taxes at a single bound. It's *superconductor!*

—ABC Evening News

The annual March meeting of the American Physical Society is like a vast, bewildering, and very long family reunion. The conferees are all physicists, and therefore share a familial fondness, but communication is difficult because they speak in dozens of mutually incomprehensible dialects of physics. So particle physicists, fluid dynamicists, and astrophysicists, for the most part, only attend talks on particle physics, fluid dynamics, and astrophysics, respectively. But at the 1987 meeting at the New York Hilton it soon became clear to the organizers that *everyone* was going to attend the hastily convened "post-deadline" session on a field that, strictly speaking, had not even existed four months earlier: high-temperature superconductivity.

Paul Grant was one of the first to see that there was going to be a problem. "I went to the APS office and told them, 'You'd better be prepared for some crowd control.' I've been to a number of rock concerts and when I was eighteen I worked the national ski patrol at a major Sierra ski resort, and I know what crowds can be when they get out of control."

On the evening of the session, March 18, physicists started gathering at around 6 P.M., two and a half hours before the first scheduled talk. The double doors to the Sutton Ballroom were locked, so an amorphous crowd began to collect in the lower-level lobby. It was like some insane physics demonstration: An escalator ran down from the main lobby, adding newcomers, one by one, to the gathering sea of scientists. The last person down was a Chinese physicist who, to avoid being swallowed by the continuously descending treads, was obliged to walk in place, hamster-like, for about five minutes until someone finally shut down the escalator. And then, at 6:45, the doors opened.

Suddenly, over three and a half thousand physicists (with twice that many elbows) seemed hellbent on proving that two bodies *could* occupy the same place at the same time. Who knew? Miracles seemed to be in the air. Two thousand of the luckiest and the fastest found seats in the ballroom. The rest spilled out into the lobbies and corridors where closed-circuit television monitors had been set up. When Cornell's Neil Ashcroft, who was the chairman of the APS's Division of Condensed Matter Physics, took to the podium and said, "Welcome to the ... *first annual* ... high-temperature superconductivity meeting," the room went wild with laughter and applause. With the suddenness of a volcanic island rising from the sea an entire new field—an entire new culture—had been born. Within a few days scientists would recall the meeting with the reverence reserved for a distant historical event. For once they had lost their heads a little, and it felt great.

Later, Michael Schlüter from Bell Labs would call it "the Woodstock of physics," and the name stuck; physicists now divide history into pre- and post-Woodstock periods. Before Woodstock, to the extent that the general public thought about physics at all, they thought of it as a science devoted to understanding the universe by disassembling it into ever smaller and, presumably, more fundamental pieces: Matter is made of atoms; atoms are made of electrons, protons, and neutrons; neutrons and protons are made of quarks . . . down and down and down, until the last, indivisible bit is discovered. Physics was Einstein and Heisenberg: the paradoxes of relativity and the epistemological strangeness of quantum theory. Books such as Gary Zukav's *The Dancing Wu Li Masters* described the pleasing ways in which the mathematical rigor of physics seemed to be blending into the subjectivity of Eastern mysticism. Particle physicists continued to discover new particles in whose fleeting lifetimes they hoped to glimpse the outlines of what they now dared to call the Theory of Everything. To pry free nature's remaining secrets the high-energy physicists in the United States were lobbying for $4 billion or $5 billion or $6 billion to build a gigantic machine known as the superconducting super collider, which would be built in Texas. (Thousands of powerful superconducting magnets, which accelerate protons to nearly the speed of light, are the essential elements of the design of the SSC. It would be ironic if the huge price tag of the SSC were to drain limited funding from fields such as condensed-matter physics that created these magnets.)

Superconductivity is a subfield of condensed-matter

physics, the study of matter in bulk—solids, liquids, and gases. Virtually all of modern-day technology owes its existence to condensed-matter physics, and this triumphant association with things practical has tainted the field, in much the same way as too much popular success will sometimes cause a writer or an artist to be unjustly criticized. The case against the intellectual purity of condensed-matter physics when compared with more "fundamental" fields has been concisely stated by the British physicist A. J. Leggett. "There are no new laws of nature to be discovered by studying condensed matter as such," he wrote, "since all behavior of such matter follows, *in principle,* from the behavior of its atomic or subatomic constituents themselves—indeed, that it is really rather a trivial occupation by comparison [emphasis added]." Carried to its logical conclusion this argument implies that the only science that has hope of uncovering any fundamental truths about the universe is particle physics. All else—superconductivity, schizophrenia, and the fluctuations of the stock market—is, in the language of textbooks, left as an exercise for the reader.

This would be merely galling if it were not for the fact that the relative obscurity of condensed-matter physics has hurt it in the war for the funding dollar and the competition for the brightest students. In the last several decades science funding has been increasingly awarded to gigantic, expensive projects: space stations, superconducting super colliders, the war on cancer, mapping the human genome. It has become increasingly difficult for fields like condensed-matter physics to draw adequate support.

Dozens of excellent popular accounts of the work of particle physicists and cosmologists have been written;

new ones seem to be published almost every day. Often these accounts are written by the physicists themselves, and by now the metaphors have been well honed. Nobody could become an elementary-particle theorist by reading one of these books, but it is possible to get a pretty good idea of the kind of things elementary-particle physicists think about. Virtually no condensed-matter physicists have taken the time to write popular accounts of their work. Philip Anderson, the acerbic and brilliant theorist from Princeton University, has his own explanation of why this is so. "It's a very hard thing to do in the first place," he says. "In the second place, we've been very busy. Particle physics goes in these great leaps and in between the leaps there is very little to do, and so you stop and you write a book either about your experiences at Los Alamos or what particle physics is all about. There are also a lot of unemployed particle physicists because they do not produce graduate students who are useful to our industrial colleagues . . . so there aren't enough of us [condensed-matter physicists] who are unemployed to do this."

Then along came Bednorz and Müller, Chu and Wu, and everything changed. All of a sudden the media abounded with news of a millennial discovery that could usher in an age of cheap energy, flying trains, tiny computers—the sky was the limit. And the cost of the crucial experiments in this new field amounted to little more than carfare for a particle-physics experiment. "I think these remarkable discoveries exemplify the importance of investment in what I would call tabletop physics," Neil Ashcroft told the thousands of cheering condensed-matter physicists who were crammed into that Hilton ballroom.

The advantages of tabletop physics were never more

clear than at the physics Woodstock. It had been less than four months since most of the world's physicists had learned about Bednorz and Müller's discovery, and sixteen days since the publication by Chu of his secret recipe for a liquid-nitrogen-temperature superconductor, but already more than fifty groups from three continents vied to present results of their latest experiments, some of which had been performed only hours earlier. "The good news," Ashcroft announced at the start of the session, "is these short reports will last no more than five minutes. The bad news is that we have the room until six in the morning." In fact, the last speaker was finished at a quarter after three.

By Friday the superconductivity meeting was on the front page of the *New York Times*. "Discoveries Bring a 'Woodstock' for Physics" the headline proclaimed, and there were pictures of Arthur Freeman from Northwestern University and Robert Cava of Bell Labs, enjoying the first of their fifteen minutes of assured celebrity. Cava was pictured holding a piece of flexible tape and a small ring. Wednesday night Cava's co-worker at Bell, Bertram Batlogg, had concluded his talk by unveiling the ring and tape to his audience of physicists. He explained that the tape would become a 90-K superconductor after it had been annealed in an oven, that applications were only a couple of years off. Then he concluded his talk by saying, "I think our life has changed." It was like Neil Armstrong saying that it was a great step for mankind, and the room went wild.

Batlogg's optimism was infectious. What if, as an IBM scientist elicited during the question-and-answer period, the flexible tape became extremely brittle once it was baked into a superconductor? So what if they could not yet carry enough current to challenge even a horseshoe magnet? Those were just technicalities, the kinds of problems that they had been trained to solve. Batlogg was right, our life had changed, and the physicists in the room were giddy with the implications. A year ago liquid-nitrogen-temperature superconductivity had seemed impossible. Now like Lewis Carroll's White Queen they were prepared to believe as many as six impossible things before breakfast. The physicists explained the economic difference between cooling superconductors with liquid helium and cooling them with liquid nitrogen with folksy comparisons: It was like the difference between scotch and water (the less affluent graduate students liked to say that liquid nitrogen was cheaper than beer).

"What it will mean," one physicist told CBS News, "is that we can do things that are completely impossible now." Another told ABC that "one could view it a little bit as the reinvention of the wheel." A year later IBM's Paul Grant was horrified when he viewed videotapes of these early news reports: "I can't believe we were saying the things we were saying." Marvin Cohen, a superconductivity theorist at Berkeley, said that he and his friends were "acting like a teenage girl who suddenly learns that she is attractive to men."

If the physicists were getting carried away, they were helped by the press, and especially by television. Michael Guillen, who has a Ph.D. in physics and teaches at Harvard, is also the science correspondent for ABC's "Good Morning America." "You feel a lot of pressure from producers and people like that to hype the story a little bit," he explains. That pressure, combined with television's voracious appetite for striking imagery, produced an exaggerated public expectation of the immediate promise of superconductivity. Every television station in the country had stock footage of the Japanese's magnetically levitated trains, which float on magnetic fields produced by low-temperature superconductors. The strong implication was that it was now only a matter of time before levitating trains would crisscross the country.

In the months following "Woodstock" the scientists did little to discourage the press; word of new discoveries, along with many false alarms, became commonplace on the front pages of the world's largest newspapers. Angelica Stacey, a chemist from Berkeley, joked about the sudden high profile of her once-invisible field. "There are new journals of physics," she told a large audience of government and industry leaders. "I think you are now reading the *New York Times* as a journal of physics. The *People's Daily of China* is also very popular . . . the next time you're standing in line in the supermarket, look at the *National Enquirer* because that's where I'm sending my work now. You can just imagine the title: Alien spaceship powered by molten superconductors."

The ease of their recent victories inspired scientists, and especially those with the cleanest hands—theorists—with the optimistic belief that all barriers could be readily surmounted. The very real and severe limitations of the new materials—their inability to carry a substantial electric current and their extreme brittleness—were hardly alluded to at all. A month after "Woodstock," after the initial rush of excitement had faded, scientists began to look at what their work had elicited from the press, and some of them were a little worried. Ted Geballe from Stanford appealed to a large audience of the new high-temperature superconductivity researchers to curtail somewhat their stratospheric flights of fancy. He feared that the cover stories in *Time* and *Business Week,* front-page newspaper articles, glitzy network news reports, and so forth had dangerously raised the expectations of the general public. If in two or three years the new superconductors had not fulfilled their promise, Geballe felt there could be severe backlash.

The press had already begun to cause some physicists embarrassment among their peers. Marvin Cohen's group at Berkeley had managed to second-guess Chu's superconductor formula. Chu had officially announced his discovery to the world at Berkeley on March 2, and at that time he had graciously acknowledged Cohen's independent, though later, discovery. Cohen got a somewhat sour taste of what it was like to be "attractive." Charles Petit, a science writer for the *San Francisco Chronicle,* interviewed Cohen and wrote a fair and accurate account of the Berkeley work. During the interview Cohen also told Petit that his group had seen hints of superconductivity at around 234 Kelvin, or minus 38 degrees Fahrenheit, which, Cohen said, was "like a cold day in Alaska." Petit was careful in his article to explain that the Berkeley group had as yet seen no more than encouraging indications of superconductivity at this high temperature, and Cohen was satisfied.

Unfortunately—and unbeknownst to Cohen—reporters write the stories, but editors write the headlines. The *Chronicle* editor thought that 234 was a more impressive number than 90, so the headline said, in bold type, that scientists at Berkeley had discovered superconductivity at 234 degrees.

The story in the *Chronicle* ran on March 3, which was also Cohen's birthday. He was awakened that morning by a phone call from a friend of his at Bell Labs. Cohen picked up the phone expecting to hear birthday greetings. "Instead he started yelling at me, 'How could you guys claim superconductivity at two hundred thirty-four degrees Kelvin?' " Cohen tried to explain that he had never made that claim, but his friend had heard it on the radio. At the "Woodstock" meeting Cohen was given a few minutes to explain, and the media virgins in the audience had their first important lesson about dealing with the press.

On July 1, 1948, the *New York Times* had devoted four and a half inches of its "News of Radio" column to the announcement of the discovery of the transistor. This short item, which was preceded by reports that WNEW Radio would broadcast traffic news on the upcoming holiday weekend and that "The Better Half" had found a commercial sponsor, is more interesting for what it did not say than for what it did. Nowhere in its 205 words does the word "breakthrough" occur. The writer also passed up such staples of science writing as "revolutionary," "remarkable," and even the ever-dependable "new." No mention was made of the imminence of a golden age or of a bright new future. It would take the next forty years of change (personal computers, electronic watches, digital sound) to drive home the point that an esoteric scientific innovation—a tiny bit of crystalline matter—could change the world. In 1987 the public was prepared to believe that a fragile, black ceramic could do it all again.

Still, the superconductivity revolution might not have seized the public imagination had it not been for a scientific parlor trick called magnetic levitation. In 1945 V. Arkadiev, a Soviet physicist at the Maxwell Laboratory of the University of Moscow, showed that a magnet would float in the space over a superconducting surface. The superconductor would act, in effect, like a magnetic mirror; the real magnet would "see" its magnetic image in the superconductor directly below it. In magnetism like repels like; the magnet hangs, like Narcissus in the thrall of his reflection, in the space above the superconductor, balanced between the downward tug of gravity and the upward repulsion from its image. To demonstrate levitation Arkadiev took a small disk of lead and a magnet and placed them in a closed cryostat of liquid helium. Since liquid helium evaporates almost instantly when exposed to normal, room-temperature air, Arkadiev could only view the levitating magnet through a small port in the side of the cryostat. It was like looking through a window into a separate and enchanted world. And this is how generations of physicists also saw magnetic levitation: a miracle too fragile to exist in their mundane world. Then Paul Chu discovered liquid-nitrogen-temperature superconductors and anyone could float a magnet in midair.

Liquid nitrogen boils at 77 Kelvin, which is minus 321 degrees Fahrenheit. It is an exquisitely clear fluid that can be stored for days in an ordinary lunch thermos. When poured over a disk of ceramic superconductor in a dish it quietly simmers while above it a cloud of white vapor forms. A magnet placed over the disk floats in the open air, a tiny miracle that is now part of our universe. The bond between scientists and magicians is ancient and powerful. For months after "Woodstock" levitation became an international scientific pastime, a trick that could finally show friends and family what superconductivity is. Most of the scientific discoveries of the last few decades have been abstract: new elementary particles that live and die in a trillionth of a trillionth of a second; the genetic code; and the Big Bang. But here was a scientific discovery that anyone could grasp, even a child.

Shortly after the APS meeting in New York Paul Grant gave his daughter Heidi, who was in the eighth grade, a thermos of liquid nitrogen, a magnet, and a superconductor to demonstrate superconductivity to her class; in just weeks a fundamental discovery had traveled from the frontiers of science to an elementary school classroom. The students were fascinated by the levitation (for that matter, they were also fascinated by the liquid nitrogen, in which flowers can be dipped and then smashed like glass). A few weeks later Grant mentioned Heidi's classroom demonstration to Paul Chu. Chu was scheduled to give a briefing on superconductivity to the National Science Board in Washington, and talking to Grant gave him an idea. Could Heidi give her demonstration to the board?

Grant loved Chu's idea and took it one step further. One weekend Heidi came into Grant's laboratory at the IBM Almaden Research Center and mixed up a batch of superconductors by herself. It was easy. She wrote up her recipe on a transparency—the "Shake 'n' Bake" method she called it, and in fact this simple way of making the 1-2-3 superconductor is generally referred to in that way. Heidi began by measuring out the proper amount of the raw materials, yttrium oxide, barium oxide, and copper oxide to make the ratio of yttrium to barium to copper, 1 to 2 to 3. She ground these materials using one of the most ancient of laboratory tools, a mortar and pestle, until she had a fine powder. She next put the powder in a heat-resistant dish and placed it in an oven at 900 degrees Celsius for a few hours. The powder fused into a dark lump in the oven, and after it cooled Heidi ground it once again in her mortar. She placed the powder into a metal die, something like a pill press, and squeezed it until it was a circular disk, embossed with the initials IBM. This disk was not yet a superconductor; it was missing too much oxygen. So Heidi once more placed the disk in an oven through which a steady stream of oxygen flowed, and baked it again. That disk, after a long, slow cooling, was a superconductor. Heidi baked twenty-five tiny superconducting cakes that weekend, enough to make presents for all the members of the National Science Board after her ten-minute levitation demonstration. Grant would call it the first practical application of high-temperature superconductivity.

The application Grant had in mind was in education. Superconductivity, he felt, was the Sputnik of the eighties, the beacon that would draw talented students into science. To explore this idea further Grant suggested an experiment to a high school chemistry teacher he knew in nearby Gilroy, California. He suggested that a group of high school students could, using only the equipment found in the average high school, make their own superconductors. The students improvised, using a pottery kiln instead of a laboratory oven and a hydraulic press from the school's shop to make the pellets. After one or two failures (the oven thermometer was not well calibrated) they were floating magnets like the pros. A group of Japanese students had already made their own superconductor, and many more student groups would follow. Superconductivity demonstration kits were soon marketed by several companies.

In fact, many people started to profit right away from the breakthrough in superconductivity. It was said that during the gold rush the only people assured of making money were those who sold prospecting equipment and maps. Now the prospecting equipment consisted of specialized pieces of laboratory apparatus, and the maps were an almost endless stream of publications. The world scientific community churned out hundreds of new scientific papers a month. A newsletter was created just to list the titles of these papers. A dozen newsletters, aimed more at industry and investors than scientists, were hastily thrown together. An investment consultant from New York City advertised that he was available to give after-dinner speeches on superconductivity, complete with magnetic levitation demonstration. An enterprising T-shirt artist at Stanford University added to his line a superconductivity shirt, on which the structure and formula of the new superconductors had been silk-screened. The comic strip "Bloom County" joked about superconductivity, and so did David Letterman (who was introduced one night as "a man whose experiments in superconductivity might someday lead to a levitating Liza Minnelli").

An entire new subculture consisting of scientists, engineers, businessmen, and science writers had suddenly sprung into existence. It even had its own faith, strongly held by some, that someday room-temperature superconductors would be possible, and that almost anyone who worked hard enough and who was clever enough could discover them. In the year following the March 1987 APS meeting several groups would announce the discovery of extremely high temperature superconductors. One of Paul Chu's announcements made it to the front page of the *New York Times*.

Optimists among those who claimed to have observed high-temperature superconductivity said that the effect was nonreproducible, implying that there had been something there, but it was too unstable and quickly deteriorated. In the past such claims would have been ignored. But now, with liquid-nitrogen-temperature superconductors a reality, even announcements of transient instances of high-temperature superconductivity had become an accepted scientific activity, or at least an activity that was tolerated.

Critics compared these fleeting observations of room-temperature superconductors to UFO sightings. Neither observation is reproducible, so orthodox science, which equates reproducibility and reality, cannot accept them. Kitazawa had even named these transient phenomena USOs, which stands for Unidentified Superconducting Objects (the word *uso* means "lie" in Japanese). Kitazawa and his colleagues have made many USO sightings at around 240 K, approximately the temperature at which other groups have also reported seeing transient evidence of superconductivity. Kitazawa soon discovered that the resistance fluctuations that experimenters had claimed as evidence for superconductivity could be reproduced at will. Just spray the samples with water, and the junction between the test leads and the superconductor then effectively functions as a battery, which plays havoc with the sensitive instrumentation.

For Kitazawa that was that, as far as the 240-K material was concerned, but not everyone shared his skepticism. Paul Chu remained a believer in room-temperature superconductivity. "There is no experimental or theoretical evidence to exclude the existence of superconductivity at much higher temperatures," he says. In fact, most new theories of superconductivity have difficulty explaining why the critical temperature is so low.

If the dream of room-temperature superconductivity and of the technology it would spawn was the carrot that drew the new field of high-temperature superconductivity forward, the stick was international competitiveness. An ABC News report in April 1987 put it most clearly: "It will probably be at least the 1990s before superconductors begin to change the ways we live, but you can get an idea of what great expectations there are even now. Guess who is already planning to capture the whole market in superconductors just as soon as they're out of the laboratory? Who else? The Japanese."

To the scientists the international race for applications was welcome only insofar as it helped attract increased funding to do experiments. Applications were interesting, but scientists were drawn more powerfully by the essential mystery of these mute, dark materials. Here was a new trick that the universe could do, which came along at a time when it was beginning to look as if the universe might be running out of new tricks. "We're facing one of the great intellectual challenges of the twentieth century," Philip Anderson says. "It is very likely that when we get the answer it will reverberate through physics for ten or twenty or fifty years. It was true of the BCS theory, which has had enormous consequences throughout all of physics: It's being used by cosmologists, particle physicists. If we understand the Big Bang it is partly because the BCS theory was discovered. Science is a seamless web."

CHAPTER 7

THE RIDDLE

The Riddle we can guess
We speedily despise—
Not anything is stale so long
As Yesterday's surprise—
                    —Emily Dickinson

On a night in early January 1987 Philip W. Anderson lay sleepless in his hotel room in Bangalore, India. Less than a month earlier, the Princeton University physicist, who is noted almost as much for his supreme and vocal self-confidence as he is for his penetrating brilliance, made what he would later admit was an "unfortunate bet." It was the kind of small wager, usually involving alcohol, that scientists enjoy. Anderson bet a friend at Bell Labs a bottle of good wine that the rumor of a new, high-temperature superconductor would eventually prove to be another false alarm. By Christmas researchers at Bell and elsewhere had accumulated enough data to convince Anderson that he had been wrong. There could be no doubt that the new materials were superconductors; the question now was, simply, why?

Several months earlier, even before he learned about

Bednorz and Müller's discovery in Zürich, Anderson had begun to suspect that the accepted theory of superconductivity, the BCS theory, did not adequately account for superconductivity in all known cases. Now, with materials that had critical temperatures as high as 40 Kelvin, he had become convinced of the inadequacy of BCS. It was the kind of challenge for which theoreticians are born: a new trick of nature, a taunt—"If you think you're so smart, explain *this!*" Anderson, sleepless on his bed in Bangalore that January night, felt equal to the challenge, and by the next morning he had sketched in mathematical shorthand the outlines for a new theory of superconductivity that he had cryptically dubbed the resonating valence bond theory, or RVB for short, which was based on ideas he had first had a decade earlier. Within a year Anderson would make the rounds of scientific meetings delivering a talk entitled "*The* Theory of High Temperature Superconductors," with the article most definite. After one such talk Vitaly L. Ginzburg, an eminent Russian theorist, was moved to remark: "I hope the RVB theory of Phil Anderson is wrong . . . because I cannot understand it."

"It really isn't my fault," said an unapologetic Anderson, "it's nature's."

After World War II, with the triumph of nuclear physics, solid state physics began to seem to many people to be a slightly disreputable pursuit without real intellectual depth. The ultimate secrets of the universe appeared to be locked up in the tiniest boxes. In principle, quantum theory, which was essentially complete by 1930, was all that was needed to understand solids. Of course, the details still remained to be worked out, but this task seemed to many to be unworthy of true Natural Philosophers. Anderson remembers feeling self-conscious about his decision to do solid state theory. But, as he told Jeremy Bernstein in his book *Three Degrees Above Zero,* after a few years he made a discovery: "The discovery was that the problems in the physics of ordinary matter . . . were just as intellectually challenging and exciting as those in particle physics . . . I suppose many of my own fellow graduate students thought I was picking an easy option, and it certainly wasn't until much later that I myself lost a certain defensiveness about the intellectual respectability of solid-state theory."

The phenomenon of superconductivity resisted all attempts at theoretical understanding for more than fifty years after its discovery. In those years, armed with the newly forged quantum theory, physicists rode roughshod across the universe. Their understanding of the nucleus led to the release of atomic energy; their understanding of solids led to the invention of the transistor and the microelectronics revolution. By the 1950s virtually all known physical phenomena evidenced by matter in bulk had, to at least some extent, been explained in terms of the quantum theory. The one flagrant exception was superconductivity; nobody really had any idea of how superconductivity worked at the microscopic level.

"Why did it take so long to explain superconductivity?" Vitaly Ginzburg once asked. "Well, it wasn't for the absence of an Einstein, because Einstein tried." In fact, most of the great theoretical physicists of the day had wrestled with superconductivity and failed to explain it. The reason, in part, was that superconductivity is what is known as a collective phenomenon, a phenomenon that results from the cooperative action of many atoms. A single atom of tin cannot be superconducting; a lump of tin can be. Examples of collective phenomena abound both in and out of physics: the stock market; a traffic jam; the melting of an ice cream cone. The theoretical apparatus devised by physicists in the first half of this century was good at handling the behavior of, say, a single electron as it raced through a metal, but was overwhelmed by effects that involved the cooperative action of a large number of electrons or atoms, the kind of cooperation that leads to superconductivity.

Nevertheless, to understand the resistance-free superconducting state it is first necessary to understand something about how resistance arises in ordinary metals. Consider a copper wire. On the atomic level the atoms that comprise this wire are stacked on top of one another in a three-dimensional array of awesome regularity. The copper atoms are bonded together in this lattice by sharing their outer electrons, which, as a consequence, are no longer associated with specific copper atoms. The result is a fixed lattice of positively charged copper ions (the copper atoms minus their outer electrons) awash in a sea of negatively charged electrons.

According to a famous theorem in quantum mechanics, electrons in a perfect crystal lattice move as if they were in a vacuum; the lattice of fixed ions does not impede or affect their motion. Only imperfections in the perfect regularity of the lattice can impede the electrons, and at any temperature above absolute zero imperfections are inevitable. The bonds between the atoms act like tiny springs, and the entire lattice vibrates like a block of Jell-O. The electrons can "sense" these vibrations, and occasionally one will collide with a vibrating atom and lose energy. It was apparent that somehow the electrons in a superconductor must not suffer from their inevitable collisions with the lattice, but how they managed this remained a mystery for many years.

The first important clue to the mechanism of superconductivity came in 1950 following the observation by Emmanuel Maxwell of the National Bureau of Standards and independently by a group at Rutgers University under Bernard Serin of what is now known as the isotope effect. Most chemical elements occur in several variant forms, or isotopes. Isotopes of a given element are chemically identical to one another, but their nuclei contain different numbers of neutrons, and so they differ slightly in their mass. The two groups of experimenters made crystals from different isotopes of the same superconducting element and measured their transition temperatures. They found that the transition temperatures of crystals made from heavy nuclei were slightly lower than those of crystals made from the lighter nuclei.

After observing the isotope effect Serin phoned John Bardeen, a theorist at the University of Illinois. Two years earlier Bardeen, who was then at Bell Labs, along with William Shockley and Walter Brattain, had invented the transistor. The significance of the isotope effect was immediately clear to Bardeen. The only difference between two superconductors made from different isotopes was the mass of the lattice; the heavier the crystal lattice the more sluggish its vibrations. Solid state physicists call these vibrations phonons, and the isotope effect told Bardeen that the more muted the phonons, the lower the transition temperature. It was clear, therefore, that somehow the interaction of the conduction electrons with these phonons was causing superconductivity. Herbert Fröhlich, a German physicist who was visiting Purdue University, had almost simultaneously, and before learning of the isotope effect, proposed that the origins of superconductivity were somehow tied to the electron-phonon interaction. Bardeen spent much of the next seven years attempting to parlay this insight into a full-blown theory of superconductivity.

In his effort to explain superconductivity Bardeen collaborated with two exceptional young physicists, Leon Cooper and Robert Schrieffer. Cooper was a native New Yorker, a graduate of the famous Bronx High School of Science, which has produced more Nobel laureates than any other high school in the world. He received his Ph.D. in physics from Columbia University in 1954 and then went to the University of Illinois for postdoctoral work. His midwestern colleague Bardeen liked to call him "my quantum mechanic from the East." Office space was tight at the University so that Cooper was forced to share a tiny office with graduate-student theorists. One of them was a student of Bardeen's, Robert Schrieffer.

Cooper had the next important insight into the nature of superconductivity. He had done theoretical calculations that indicated that under certain conditions the electrons in a metal could form pairs. This was remarkable because like charges repel; in ordinary metals electrons avoid one another. But Cooper saw that another, very weak force also acted between electrons, causing them to form weakly bound pairs, which his intuition told him would lead to superconductivity.

An electron as it speeds through a crystal lattice can be

likened to (among a great number of things) a car racing down a highway. As it speeds along the car cleaves the air in front of it; trailing behind the car, where it had been only a moment before, is a vacuum, a vacancy in the atmosphere quickly filled by inrushing air. A tailgating car would be drawn, along with the returning air, into this vacuum. The rear car is, effectively, attracted to the one in front. Racing-car drivers call this effect "slipstreaming," and often exploit it to great advantage. Electrons in a metal can attract each other in much the same way. A negatively charged electron racing through a crystal draws the surrounding ions—the positively charged atomic cores—toward it. The electron races on, but it takes another instant for the lattice to return to its relaxed state. In the meantime the puckered region in the lattice contains a slightly higher concentration of positive charges than its surroundings. An electron passing nearby will be attracted to this positive pucker in much the same way that a tailgater is drawn forward by the leading car.

Cooper saw that the weak attraction between electrons in a solid could cause them to form loosely bound pairs (which are now known as Cooper pairs). This insight did not immediately lead to a theory of superconductivity. Serious quantum mechanical issues still remained to be solved, and Bardeen suggested to his student Schrieffer that this would be a worthy thesis problem. Schrieffer agreed and spent many fruitless months working on the problem. By the end of 1956 he was ready to give up on superconductivity and find some easier topic for his thesis. He discussed this with Bardeen, who urged him to work on it for another month before quitting. In the meantime, Bardeen had to go to Stockholm to accept the 1956 Nobel

Prize in physics, which he shared with Shockley and Brattain for their invention of the transistor.

In Bardeen's absence the two younger physicists found what would prove to be the final piece of the puzzle. The final explanation of superconductivity is quantum mechanical; the superconducting electrons in a metal behave as if they were part of a gigantic atom, and just as atomic electrons remain forever in their atomic orbitals, superconducting electrons can circulate in a perpetual current.

After Bardeen had returned from Stockholm the three physicists quickly assembled the ingredients into a complete theory of superconductivity that not only explained all known experiments, but predicted new phenomena (but no new superconductors). They wrote two now-classic papers in which their theory was described in detail, which were published in *The Physical Review*. The first public airing of the theory, however, was given at the March meeting of the American Physical Society in 1957.

The beauty of a physical theory is often almost as important as whether it agrees with experiment; experiments, after all, can be wrong. Philip Anderson has observed that "a theory which contradicts some of the accepted principles [of physics] and agrees with experiment is usually wrong; one which is consistent with them but disagrees with experiment is often not wrong, for we often find that experimental results change, and then the results fit the theory." The BCS theory was absolutely gorgeous. Niels Bohr told Schrieffer that he thought the theory was absolutely beautiful in its simplicity; his only worry was that it might be too simple to be true. By 1959 the BCS theory was already so widely accepted that at a conference David Shoenberg from Cambridge University introduced an experimental talk, only half joking, by saying "let us see to what extent the experiments fit the facts."

The BCS theory is one of the supreme triumphs of twentieth-century science, but not everyone was equally enamored of it. BCS was very good at describing superconductors that had already been discovered, but was no good at all at predicting which materials would become superconducting and at what temperature. Bernd Matthias seized upon this shortcoming, deriding theorists every chance he got. "Any real understanding always leads to a procedure which will enable us later on to predict," Matthias insisted. "Only once a problem is understood properly can it be predicted, and therefore prediction is an extremely good criterion. The people who don't seem to be able to predict anything may also not understand anything."

One important prediction that BCS did make was an upper limit on the possible critical temperature. Unfortunately, it was the kind of prediction that could only be proved in the negative, by finding a superconducting material of higher critical temperature. The theoretical reason for the limit, which Philip Anderson and Marvin Cohen calculated in 1972, arises from the observation that above the critical temperature the Cooper pairs are destroyed by the thermal vibrations of the crystal lattice, which increase with increasing temperature. It follows that the stronger the coupling between the lattice and the electrons the higher the critical temperature would be. But before the critical temperature gets very high the coupling between the electrons and the phonons literally rips the lattice apart, forcing the atoms to rearrange themselves into another, more weakly coupled configuration. Just how high

T_c could go before this happened was difficult to calculate, but no matter how it was calculated it could not be much greater than 40 Kelvin, which, from the point of view of technology, was extremely frustrating.

To overcome this barrier theorists began to cast about for other or "novel" superconducting mechanisms. In 1960, William Little, a British theorist newly arrived at Stanford University, had a wild idea for a new mechanism of superconductivity, an idea so crazy that it had a chance of being right.

Before coming to Stanford Little had completed two separate doctoral degrees, from the University of South Africa and from the University of Glasgow in Scotland. Little was intrigued by an observation he read about in an early book on superconductivity by Fritz London, who had developed a macroscopic theory of superconductivity in 1950. London recognized, even before the BCS theory, that the motion of the electrons in a superconductor was highly organized. This organization reminded London of the type of organization that arises in biological systems. Little had also been impressed by this resemblance, and by the stability of the superconducting state in the face of external influences, such as heat. "It occurred to me," Little wrote, "that if nature wanted to protect the information contained, say, in the genetic code of a species against the ravages of heat and other external influences, the principle of superconductivity would be well suited for the purpose." London died before the BCS theory, so he was unable to investigate his ideas further. After reading London's book Little, armed with the new BCS theory, decided to try a calculation.

For his prototypical molecule Little chose a highly simplified version of DNA, essentially just a long chain of carbon atoms. To this chain Little imagined affixing at regular intervals shorter molecules that projected to the side like the oars of a galley. The side chains would be constructed to behave a lot like bits of paper that are attracted to an electrostatically charged pocket comb. The charge of the comb pushes some of the electrons in the paper away, unmasking the underlying positively charged ions. The positive and negative charges on the surface of the paper become slightly separated; the paper becomes polarized. The positively charged side of the paper is attracted to the negatively charged comb. An electron racing down the molecule's long spine polarizes the side chains, just as if they were bits of paper. The side chains remain polarized for a short amount of time after the electron passes; the next electron to race down the spine is attracted to the freshly polarized side chain. In effect, the first electron attracts the second. This attraction, Little conjectured, could lead to the pairing of electrons that is the prerequisite of superconductivity. Little found that this mechanism should give critical temperatures of as high as 1000 Kelvin, or 1340 degrees Fahrenheit: superconductivity at room temperature on Mercury!

Little's hope was that he could design superconducting molecules that organic chemists could synthesize. Unfortunately, this proved to be very difficult in practice. Many physicists scoffed at the idea that an organic compound (that is, a carbon-based compound) could become metallic, let alone superconducting. More than a dozen organic superconductors have now been discovered, but none have very impressive critical temperatures—and all of these could be entirely explained by the BCS theory. Organic compounds of the type Little envisioned are very difficult to make, so the trial-and-error search for organic superconductors is a slow and tedious affair.

Little's theory—the excitonic theory—found few strong advocates in the United States. Matthias, true to form, ridiculed Little's idea both publicly and mercilessly. Then, in the last years of his life, he abruptly changed his mind, saying that superconductors of the sort suggested by Little would one day be discovered. Little was glad to be spared Matthias's barbs, but was not particularly heartened by winning his conversion: "I had as much confidence in Matthias's latest predictions as I had in his previous predictions, because they all appeared to be based on the absence of knowledge. I think he was just being friendly." The Russians were greatly impressed by Little's theory. The strength of Russian science is theory, and the excitonic mechanism was a theorist's dream, a kind of superconductivity construction set. The excitonic mechanism has since had a deep influence on Russian theoretical work.

A handful of other superconducting mechanisms have been proposed by various scientists around the world. But before Müller and Bednorz's breakthrough, none had seemed to describe actual superconductors. Then, post–Bednorz and Müller, theorists rushed to their closets, dusting off old theories. It was like the Cinderella story, with the scientists attempting to fit their old feet into this remarkable new slipper.

But on that January 1987 night in Bangalore, Phil Anderson had come up with an essentially new theory of superconductivity. Anderson's intuition was sparked by the pictures of the lattice of Bednorz and Müller's materials provided to him by friends from Bell Labs.

An essential feature of all the known high-temperature

superconductors—the ones known to Anderson at the beginning of 1987, and those discovered since—is the existence of planes of copper and oxygen. The copper and oxygen atoms are arrayed on a square grid; the coppers lie at the intersections of the grid and there is one oxygen between neighboring pairs of coppers. Between these planes are layers consisting of various other atoms—yttrium and barium in the case of the original family of liquid-nitrogen-temperature superconductors. This configuration is technically known as the perovskite structure. Perovskites can be constructed out of a wide variety of different elements and display a remarkable array of electrical properties. Some are insulators, some are conductors. Some, when squeezed, generate tiny electric currents. Some are ferroelectrics, the electrical equivalents to permanent magnets. Even before Bednorz and Müller's discovery showed that some perovskites were superconductors over $20 billion a year was spent on these useful materials.

After staring for a while at the structural diagrams of the new superconductor Anderson decided that the electrons in the copper-oxygen plane, and particularly those that were restricted to the coppers, were the perpetrators of the superconducting state. The magnetic properties of electrons, not their electrical properties, are crucial to Anderson's resonating valence bond theory. Electrons can be thought of as tiny, spinning, electrically charged balls. The spinning charge generates a magnetic field, and so the electron acts as if a bar magnet were oriented along the spin axis. In Anderson's model the pairing of electrons is a magnetic effect; spins are magnetically linked to other spins.

Anderson was by no means the only theorist racing to discover a theory that would account for these new superconductors. And almost daily the experimentalists provided them with new data to explain. Several groups discovered that there was no isotope effect in 1–2–3, the liquid-nitrogen-temperature superconductor found by Paul Chu, and this seemed to rule out the phonon mechanism and the standard BCS theory. The entire analytic apparatus of late-twentieth-century physics was brought to bear on these black ceramics. The new superconductors were scrutinized with beams of X rays, electrons, neutrons, positrons; with powerful magnetic fields; and with infrared light and lasers. It is doubtful that so much has ever been learned about a single compound in so brief a time. But to a large extent theory determines what is seen. So far no theory, including Anderson's, has been developed far enough to make a unique prediction, and only a very few theories have been definitely eliminated. Research into the new superconductors will have to proceed without theoretical guidance for a while.

Like the BCS theory, the theory of the new superconductors, when it is discovered, may not guide scientists to the discovery of new materials. But BCS theory has had many spinoffs in other branches of physics, especially in the theory of elementary particles and in cosmology. Knowledge of the tricks of which nature is capable can have an importance that goes beyond their technological applicability. "A theory is more," Leon Cooper said in his 1971 Nobel Prize acceptance address. "It is an ordering of experience that both makes experience meaningful and is a pleasure in its own right."

CHAPTER 8

SUPER-TECH

In the nineteenth century physicists had worked and schemed to achieve ever lower temperatures, to approach the quiescent purity of absolute zero, where they presumed atomic motion would cease and the laws of physics would bare themselves to inspection. This quest led, among much else, to the discovery of superconductivity. Using the technology based on this discovery, twentieth-century physicists are planning to build a vast machine with which they hope to approach another absolute, the absolute zero of time, where their theories tell them the essential unity of the universe will be manifest.

The machine, which is called the superconducting super collider (SSC), gives new meaning to the term "Big Science." It will consist of an oval ring 52 miles around, precise in all of its dimensions to within a thousandth of an inch. Two beams of protons, guided by magnets, will race around this giant ring, in opposite directions, at very nearly the speed of light before they are steered toward head-on collisions that would, in effect, roll back time to the brink of creation, fleetingly re-creating the state of the universe that reigned one-millionth of a billionth of a second (10^{-16} second) after the Big Bang. In these fiery explosions of ur-matter physicists hope to find the missing clues that will lead them to the realization of Einstein's lifelong dream of uncovering the Law within the laws, a unified theory of space and time, matter and energy, the Theory of Everything.

Particle physicists, who customarily work with the most avant-garde technology, were among the first to employ superconductors. In fact, without superconductivity much of the progress toward a Theory of Everything that particle physicists have made over the past twenty-five years would not have been possible. Superconducting magnets, which warp the trajectories of speeding charged particles, are key ingredients of most of the detectors used to analyze subnuclear collisions. The world's most powerful accelerator, the Tevatron, located at Fermilab in Illinois, is a ring of superconducting magnets 3.5 miles around. The SSC will be a scaled-up version of the Tevatron, a 52-mile-long necklace of over 10,000 superconducting magnets. Large as the SSC will be with these superconducting magnets, it would be unthinkably immense without them. Built from conventional copper and iron water-cooled magnets, the SSC ring would be about 180 miles in diameter. The electric bill to run these magnets would be about $2 billion a year, compared to an estimated $50 million a year for the superconducting machine.

By the spring of 1987, funding to begin construction of the SSC seemed assured. It had been a long and sometimes bitter battle. Many of the opponents of the SSC were tabletop physicists who believed that the scientific insights that the SSC might someday provide did not justify its expense, and that it and other big-ticket projects drain money from many worthy and important areas of research. They noted that the SSC was just a monumentally scaled-up version of existing accelerators and suggested that perhaps particle physicists should concentrate on developing better and cheaper ways for accelerating particles. After all, what was the rush? The universe was not going anywhere. The particle physicists countered that without the SSC a whole generation of talented students would turn to other fields, and without them the field would wither. The debate went back and forth for years. Then, on the eve of victory, with presidential approval of the costly project finally obtained, high-temperature superconductors became front-page news, and for a while it seemed as if the designers of the SSC would be sent back to their drawing boards.

On April 14, 1987, the *New York Times* ran an article opposite its editorial page by Philip Anderson, who is a scornful critic of Big Science, and especially of the SSC. "It is important to wait a while on the SSC," he wrote. Preliminary measurements indicated that the new superconductors could someday generate magnetic fields ten times or more stronger than possible with the low-temperature superconductors. With magnets ten times more powerful, Anderson argued, the SSC could be one-tenth as large, saving billions of dollars. The smaller machine could be built on a site that is one-hundredth the size of the currently proposed site, which would be a substantial saving. And even if the high-temperature superconducting magnets were no stronger than the low-temperature superconducting magnets, they should be much less expensive to operate because they could be cooled with liquid nitrogen instead of costly liquid helium. It was as if the particle physicists had been planning to serve champagne at their extravagant party, and Anderson was telling them that beer would do. In his *Times* article Anderson admitted that originally he had thought that "technology was not going to move fast enough to make a difference to the supercollider and I've changed my mind. Things are moving faster than I ever thought."

Official Washington had begun to learn about the dramatic advances in superconductivity in December 1986 as scientists who had attended the Boston Materials Research Society meeting reported in to their program managers at the government agencies from which they received funding, requesting permission to divert monies to research in high-temperature superconductivity. Within a few months federal research agencies rechanneled $45 million of their fiscal 1987 funds to research in high-temperature superconductivity. In the meantime, congressional science-watchers, who had fewer direct connections with working scientists, read about Bednorz and Müller's discovery in *Science* and *Nature,* and were intrigued. "Our guys said, hey, this is exciting, this is interesting," recalls Paul Maxwell, a materials scientist attached to the House Committee on Science and Technology. "Then we heard about the Chu thing," Maxwell remembers, "and that was the dam breaker."

The House Committee on Science and Technology, together with the science and technology subcommittee of the Senate Committee on Commerce, Science and Technology, is charged with the task of apportioning most of the federal research dollars among the various government agencies, such as the Department of Energy, which funds most particle-physics research. In addition to the SSC the committee is responsible for funding the space shuttle and the space station, and had been directed by the president to double the budget of the National Science Foundation by 1992. Even before Anderson's article had appeared in the *Times* House members had begun to wonder whether it might be a good idea to wait a bit on the SSC and see if the new materials might mean a significant savings. As early as February 1987, only four days after Paul Chu had announced the discovery of liquid-nitrogen-temperature superconductors to the world, but weeks before he divulged the correct composition, James Krumhansl, solid state physicist from Cornell and president of the American Physical Society, sent a letter to Energy Secretary John S. Herrington: "The implications [of high-temperature superconductivity] are vast! These materials are inexpensive, they can be easily made, and can be refrigerated by a variety of widely available, cheap methods. They unquestionably have the potential to save billions of dollars in construction and operation of particle accelerators like the SSC. Because they are easily fabricated, I have little hesitation in predicting that they will be brought to technological usability in three to five years, if materials research is supported adequately. By contrast with particle physics, I can assure you that this discovery is so important that it will find its way into almost every area of materials, energy, electronic and military technologies. A scientific development such as these new materials comes once in decades."

Backed by the authority of at least two distinguished scientists, congressional opponents of the SSC took up the issue of high-temperature superconductivity. Buddy MacKay, a Florida Democrat, and Don Ritter, a Pennsylvania Republican and the only member of Congress who holds a doctor of science degree, circulated a letter in which they argued that it was worth reconsidering the design of the SSC in light of the recent breakthroughs. It might turn out that billions of dollars could be saved without hindering the progress of particle physics, money that could be spent on fields of scientific research that are now relatively neglected.

But proponents of the SSC noted that it had taken some twenty years to learn how to build magnets to the demanding specifications required by the SSC. Barring a miracle, Krumhansl's optimistic three to five years would probably stretch out to ten or more, and such a delay would lead to the loss of an entire generation of particle physicists. In a passionate letter to the *New York Times,* three eminent particle physicists, Howard Georgi, Sheldon Glashow, and Kenneth Lane, wrote, "There is great potential for the new superconductors. They may lead to swift and silent subways and trains, and to efficient power transmission . . . They could even lead to a renaissance of American industry. They should not be used to delay, defer and derail American high-energy physics."

A careful analysis by the designers of the SSC indicated that Anderson and Krumhansl had probably overestimated the savings possible by using the new materials. It is true that stronger magnets can steer protons into tighter orbits, but the stronger magnetic fields lead to other problems. For instance, the coils themselves would be subjected to powerful forces by their own magnetic fields. Designing magnets that could withstand these forces was one of the most challenging problems for the architects of the SSC; magnets ten times stronger, and especially made of brittle ceramic superconductors, might be impossible or prohibitively expensive to build. According to Leon Lederman, director of Fermilab, replacing the SSC magnets with high-temperature superconducting magnets of equal strength would save only about 5 or 10 percent of the uninflated cost of the super collider because of reduced refrigeration costs—too little to justify a five-year delay. Besides, technology, by its very nature, is constantly improving; waiting for the *best* videotape recorder, microwave oven—or super collider—can mean never owning one at all.

The particle physicists, by dint of their eloquence and unity, as well as long experience in Washington, won this round; the federal government decided not to delay the construction of the SSC. But, as this story illustrates, high-temperature superconductivity had, in less than two months, gone from being an esoteric scientific discovery to being a political issue. The promise of almost limitless applications, of course, had much to do with the speed of this transformation. But it also reflected the growing political importance of science. This trend is described by David Dickson in his book *The New Politics of Science.* Dickson notes that ever since World War II, "advanced technology has become the key to economic and military power. Over the same period science has become the key to advanced technology ... The last few years have seen the reemergence in the U.S. of an almost religious belief—dormant for much of the 1970s—in the powers of science-based technology." And perhaps the most fervent believer was Ronald Reagan.

On July 22, 1987, Ronald Reagan emerged from a one-hour White House briefing bursting with excitement. "I've just spent the last hour learning all about recent discoveries in the field of superconductivity," he told a group of visitors. "Considering the kinds of teachers I've had, I should be an expert on the subject. So fire any questions about superconductivity you have and I'll do my best to answer."

Reagan's briefing on superconductors came less than five months after Wu and his graduate student had removed the first impure, greenish ceramic superconductor from the oven in their laboratory at the University of Alabama. In those few months the new superconductors had been subjected to virtually every kind of scrutiny known to science. There had been no "deal-breakers," no insurmountable problems that stood in the way of eventual application. The toughest questions—when would the first products containing superconductors reach the market, and who would make them—lay outside the analytical realm of science. But scientists and industrial leaders who had visited Japan returned impressed with the magnitude of the Japanese effort to be the first to commercialize high-temperature superconductivity. Japanese companies had already filed close to a thousand patents for superconducting devices and processes using the new materials. The formidable Ministry of International Trade and Industry (MITI) had reportedly started assembling one of its unfailingly efficient consortia. Articles on superconductivity filled Japanese newspapers; books on superconductivity were instant best-sellers. The message was clear: Japan, which sells the United States all of its VCRs, most of its computer chips, and many of its cars, was now determined to supply it with superconductors as well.

On June 10 the House Science and Technology Committee conducted a hearing on high-temperature superconductivity. Paul Chu testified, as did leading scientists from IBM and Bell Labs. Senators David Durenberger (R-Minnesota) and Pete Domenici (R-New Mexico) and Congressman Don Ritter, all of whom would soon introduce legislation to speed the development of superconductivity, testified. The nine hours of testimony were recorded by TV cameras from all the U.S. networks, and from Japanese television. The walls of the hearing room were covered with color enlargements of the space shuttle and of astronauts floating high above the earth, reminders of a time when the United States was the world's unquestioned technological leader. But the presence at the hearing of the Japanese was a reminder that for more than a decade, despite its continuing scientific excellence, the United States' once-unquestioned technological preeminence had been slowly waning. Superconductivity was viewed by many in Washington as a last chance for America to show that it can compete effectively with the Japanese. A report prepared for Congress by the Office of Technology Assessment starkly summed up the challenge: "If Japan were to surpass the United States in a new science-based technology like high temperature superconductivity, U.S. competitiveness could be very broadly threatened. The stakes have quickly come to seem a good deal greater than superconductivity itself."

On July 28, not quite a week after his White House briefing, Reagan displayed his newly acquired knowledge before an audience of 1200 of America's leading scientists, business executives, and policymakers, who had gathered in Washington for the Federal Conference on the Commercial Applications of Superconductivity, a two-day meeting-*cum*-pep-rally that had been hastily arranged by the White House Office of Science and Technology Policy. To underscore the importance of the occasion (and, perhaps, to divert attention from Ed Meese's testimony before the Iran-Contra committee) Reagan had brought along with him three ranking Cabinet members: Secretary of State George Shultz, Secretary of Defense Caspar Weinberger, and Secretary of Energy John Herrington. "Science tells us that the breakthroughs in superconductivity bring us to a threshold of a new age," Reagan said. "It is our task at this conference to herald in that new age with a rush . . . It's our business to discover ways to turn our dreams into history as quickly as possible. The laboratory breakthroughs into high temperature superconductivity are an historic achievement. But for the promise of superconductivity to become real, it must bridge the gap from the laboratory to the marketplace. It must make the transition from a scientific phenomenon to an everyday reality, from a specialty to a commodity."

To speed the transition Reagan announced an eleven-point Superconductivity Initiative (which prompted the *New York Times* to observe that the word "initiative" had become "the official way to describe any program that is still more goal than fact"). The president's Initiative featured the formation of what he called a "Wise Man" Advisory Group on Superconductivity, the establishment of a Superconductivity Research Center, and changes in antitrust and patent laws to promote innovation. He also called for the Department of Defense to spend $150 million over the next three years "to insure the use of superconductivity in military systems." Reagan promised that his administration would soon be announcing its own legislation, which would vie with the half a dozen acts already proposed by House and Senate members from both parties.

The United States government spends about $60 billion a year on scientific research and development, which is approximately the same amount that industry spends. This money is doled out by Congress to the national laboratories and the so-called "mission agencies," the Department of Defense, the Department of Energy, the National Science Foundation, and others, who in turn fund research that furthers their missions. Much of the generic technological information this money generates is available to whoever is interested in it whether in the United States or abroad. From time to time some of the technology that the agencies develop in the pursuit of their missions is of such general use that it is spun off. But, unlike many other nations, the United States, in accordance with the economic theories of the Reagan administration, does not directly support commercial research and development, fearing that such measures would distort the functioning of the free market. Instead, the United States attempts to stimulate commercial innovation by indirect means, for example, special R&D tax credits. In his Superconductivity Initiative, Reagan proposed changes in antitrust laws to encourage cooperative research and improvements in patent laws to provide better protection of so-called intellectual property.

In fact, superconductivity—low-temperature superconductivity, that is—provides a good example of how government-funded research sometimes leads to important commercial innovation.

The magnetic resonance imaging scanner, or MRI, represents the most significant advance in medical imaging since the invention of the X-ray machine by Wilhelm Röntgen in the last years of the nineteenth century (for which Röntgen was awarded the first Nobel Prize in physics, in 1901). MRI provides pictures of the human body of unprecedented clarity and detail. MRI can make remarkably detailed maps of soft tissue and of veins and arteries, which would be virtually invisible to X rays without injection of special contrast agents. "With CT [computerized tomography, an X-ray technique] scan, we mapped the interstices of the human body," one physician says. "With MRI we'll see the back alleys." An added bonus of MRI is that it creates its detailed three-dimensional image without exposing the patient to radiation. The patient is exposed only to a strong magnetic field, which is harmless (as long as the patient does not have a pacemaker or any other metallic implants). The powerful magnetic field in most commercially available MRI systems is generated by a large superconducting coil.

Almost immediately after he discovered superconductivity, Kamerlingh Onnes began to dream up practical applications for this strange, new phenomenon. He did not have to tax his imagination unduly; energy loss is the albatross of engineering, and electrical resistance is one of the primary reasons for energy loss. Large electromagnets were useful laboratory tools in Onnes's day, as they are now. Almost all the energy lost in operating a giant electromagnet is due to electrical resistance, and the loss takes the form of heat. A powerful superconducting magnet could be far smaller than a resistive magnet because its windings could carry large currents with no loss. Better still, once the current is set racing through the coil it will continue its circulation forever; the power supply can be removed. The only cost would be the cost of keeping the superconductor cold. Within a year or so of his discovery Onnes tried to make a superconducting magnet.

Onnes found that the frail superconducting state was destroyed by even a small magnetic field or a trickle of electric current. The engineer in him was disappointed, even as the scientist rejoiced in a new discovery. Nonetheless, Onnes remained confident that these problems were not inherent to all superconductors, and would in the future be solved. He wrote in 1913, "As we may trust in an accelerated development of experimental science this future ought not to be far away."

In fact, that future would not dawn until the early 1960s, when Eugene Kunzler at Bell Labs won twenty-one bottles of scotch by proving that some superconductors could tolerate a strong magnetic field while carrying a large electric current. Kunzler's discovery was only the first step toward the ultimate construction of reliable, high-field superconducting magnets. These new superconductors required special processing techniques before they could be made into suitable wires. Materials scientists needed another decade of concentrated effort to learn enough about these superconductors to be able to draw them reliably into wires from which magnets could be wound.

The expensive development of superconducting magnets would not have been undertaken unless there had been customers for them. Fortunately, elementary-particle physicists—who are principally funded by the Department of Energy—had already mastered the art of building large superconducting magnets, and they were happy to share this information. One of the first workers to attempt to build a full-body NMR scanner (the initials MRI would come later, probably because the word "nuclear" in NMR —nuclear magnetic resonance—alarmed some patients) was a young doctor named Raymond Damadian. Damadian did not know the first thing about superconducting magnets when he began his quest to build his first imaging scanner, a machine he called Indomitable. So he called the scientists at Brookhaven National Laboratory who had designed superconducting magnets to be used in an accelerator named ISABELLE that was never built. The Brookhaven scientists happily gave Damadian a copy of a computer program they had written to design superconducting magnets, and, thanks to government funding of pure research, Damadian was in business.

Damadian succeeded in demonstrating the technique of magnetic resonance imaging, but it took more experienced engineers to develop the MRI scanner into a successful commercial product. The British company EMI, which had been instrumental in the development of the CAT scanner, built the first prototype MRI machine in 1978, followed the next year by a West German company, Bruker. These companies were also free to draw on the considerable store of U.S.-funded superconductivity R&D. Commercial MRI systems became available in 1984.

MRI remains the only commercial success story in the history of superconductivity. By 1987 over 1300 MRI systems, costing from $1.5 million to $2.5 million each, had been installed in hospitals around the world; over 500 of them had been sold in that year alone, by two dozen companies, which works out to industry sales of about $1 billion. The market for MRI machines is expected to increase by 20 percent for 1988. High-temperature superconductors operating at liquid-nitrogen temperature could reduce the price, size, and operating costs of MRI systems, and make them easier to operate and maintain. According to estimates by Lawrence Crooks of the University of California, San Francisco, the price of an MRI system might be reduced by about 8 percent. That may not seem like very much, but to the Japanese, who must import all their helium, nitrogen-cooled superconductors are especially attractive.

MRI is one example of the way the government's mission agencies spin off technology: The Department of Energy supported the pre-commercial development of superconducting magnets. But seventy cents of every dollar spent by the federal government on R&D goes to national defense. At one time, roughly from the end of World War II to the Vietnam War, money spent on defense-related R&D helped to develop computers, semiconductors, and other commercially important technologies. But in recent years the needs of the defense sector have been for the kind of specialized and sophisticated technology that has few civilian uses. It is a trend that could have serious consequences for the commercialization of superconductivity given that nearly half the R&D funding for high-temperature superconductivity proposed by the government will go to the Department of Defense.

While DoD is unable to do the kind of research needed to lead to the commercialization of high-temperature superconductivity, it seems that private industry is unwilling to do it. In a study highly critical of the federal policy toward the commercialization of high-temperature superconductivity that was published in the summer of 1988, the congressional Office of Technology Assessment (OTA) notes that all but a few of the largest American corporations, exempting such companies as IBM, AT&T, and Du Pont, are taking a wait-and-see attitude toward superconductivity: "They plan to take advantage of developments as they emerge from the laboratory—someone else's laboratory—or buy into emerging markets when the time is right. Unfortunately, reactive strategies such as these have seldom worked in industries like electronics over the past 10 to 15 years, while many American firms seem to have forgotten how to adapt technologies originating elsewhere."

The deregulation of the financial markets during the Reagan years has led to an increasing emphasis by corporate planners on investments that yield short-term gains. According to the OTA, "Few managers view research as a major element in long-term competitive strategy." Therefore, the government needs to foot the bill for the preliminary, basic research that industry cannot, or will not, do. This work is, for the most part, carried out at America's universities. Most of the support for university science in the United States—which has long been considered its greatest strength, and which also educates future generations of scientists—is provided by the National Science Foundation. An inspection of the breakdown of R&D money among the various agencies shows that the NSF has not fared as well as other agencies. In 1987 the NSF and the DoE spent approximately equal amounts on high-temperature superconductivity; in 1988 DoE's budget more than doubled, while NSF's increase was less than 25 percent. The NSF had to turn away many highly rated proposals in high-temperature superconductivity. The ten DoE laboratories got more money to spend on high-temperature superconductivity in 1988 than the nation's universities got from the NSF. The OTA report concludes from this that "the allocation of Federal R&D funds seems out of balance, given the great strength of American universities in basic research."

After the stock market collapse on October 19, 1987—Black Monday—administration officials and congressional leaders from both parties agreed that something finally had to be done to begin paying off the nation's massive $2.5 trillion debt, more than half of which had accumulated since 1981. In the last months of 1987 Congress produced a revised budget designed to reduce the deficit by $33 billion. The budget allowed $67.1 billion for R&D, $6 billion less than the administration had requested. In terms of constant dollars, adjusted for inflation, the R&D budget declined by 0.3 percent. The NSF had been slated to receive a hefty 16.7 percent increase, or $238 million over its 1987 level, as the Reagan administration's announced intention of doubling the NSF budget in the next five years. That increase was trimmed back by $202 million. NASA and DoD also were substantial losers. The increase in high-temperature superconductivity spending does not, therefore, represent new funding so much as funding that comes at the expense of other, less fashionable research. Not long ago, superconductivity was one of these less fashionable fields.

"Prediction is difficult," Yogi Berra is said to have once remarked, "especially of the future." Nevertheless, predicting the future can be an irresistible and invigorating sport, especially in those rare instances when the outlook is mostly bright, as it is for high-temperature superconductivity. And because superconductors had been around for a long time before the liquid-nitrogen barrier was broken, predictions were easy to make. Engineers had designed and built prototypes of all kinds of tantalizing devices, but the necessity of cooling them with liquid helium made them impractical. If, as seems likely, the new superconductors can be used in place of the old superconductors, many of these applications will suddenly become practical as well as possible, and the consequences—cheaper energy, faster trains and computers, and the like—will be enormous.

Viewing the high-temperature superconductors as replacements for the low-temperature ones leads quickly to predictions. The transistor, in the first years following its invention, was also seen as a replacement technology. In this view the transistor was just like a vacuum tube, only it was smaller, and it did not heat up as much or consume as much power. And the first devices containing transistors, such as portable radios, did use transistors in just this way. It was only after a generation of engineers had become used to thinking in terms of transistors that new inventions emerged that would have been impossible with vacuum tubes—things like digital watches and personal computers. Ideas like these come as much from the collective imagination of an age steeped in a new technology as from any individual's imagination. This phenomenon has given birth to one of the enduring clichés of the superconducting age: The most important applications will be those that cannot yet be anticipated. Nevertheless, it is fun to try.

Stanford University's Ted Geballe, who has spent his long career thinking about superconductivity, described the pitfalls of prognostication at a meeting at Fermilab in May 1987. "Supposing we had gone from the Stone Age to the Bronze Age to the Plastic Age, and for some reason we bypassed the Lodestone Age and the Iron Age, and all of a sudden last year, Dr. Müller over in Zürich discovered iron," Geballe asked. If he was then asked to predict the future of magnets, what would he come up with? "Well, if I want to hang a memo on my file cabinet, I get one of those nice little pieces of [magnetized] iron and it will hold the memo right there where I want it to be. That's an example of non-imaginative thinking."

As long as he was giving his imagination rein, Geballe started by assuming that "we're going to have room-temperature superconductors to play with in the twenty-first century." First, it seems obvious that room-temperature superconductivity would drastically change how civilization generates and uses energy. Electrical energy would become much more important. It could be harvested by solar collectors in the desert and transmitted with no loss to consumers in darker climes. Electrical energy could be stored in superconducting "tanks"; huge tanks could collect solar energy by day and distribute it at night. Smaller tanks could be located in private homes.

Energy tanks could also be used to power electric cars, which might levitate on magnetic fields above roadways or be powered by tiny, extremely powerful and efficient electric motors. These same tiny electric motors could be the muscles of robots, which might have superconducting supercomputers as their "brains."

Extremely sensitive superconducting sensors could be

used to map the interior of the Earth and locate valuable lodes of untapped resources. These same detectors could be used to map electrical signals that emanate from the human body every time a neuron fires. "These signals could be collected, correlated, and used for diagnosis in ways that boggle the mind," Geballe said.

It's irresistible to join Geballe on his flight of fancy. Superconducting motors could be used to make bionic arms and legs or power artificial hearts. Superconducting sensors could directly pick up signals from the brain to control these limbs—or for that matter cars, computers, or anything else. How about superconducting skis or skates? Or a superconducting mattress? Once you start it's hard to stop. Such fanciful thinking may seem like building castles in the air. Then again, superconductors might change the architecture of the twenty-first century so radically that even these will become possible.

At the conclusion of his speculative spree Geballe admitted that he had probably only "done the equivalent of saying warm superconductors will have an impact equal to a magnet holding a piece of paper on a file cabinet." Perhaps it is wiser to look at the applications of superconductivity that have already been explored and that will probably be part of the foreseeable future. What they may lack in romance they make up for in reality.

During World War II, long before the materials that would make practical superconducting magnets possible had been found, military scientists investigated applications of superconductors that exploited properties of the superconducting state other than its zero resistance. Superconductors are extremely sensitive to heat, so they can form the basis of extremely sensitive infrared sensors called superconducting bolometers, which can detect troops and vehicles in pitch darkness.

Electric motors and generators made from superconductors can be much smaller and lighter than their conventional counterparts. The navy began to develop "superconductive electric propulsion systems" for their ships in 1969. The system consists of a superconducting generator powered by a conventional ship engine, which powers a superconducting motor that is connected to the propeller. The advantage of this system is flexibility and fuel efficiency. Ordinarily a ship's engine is connected to the propeller through a complicated and bulky mechanical transmission; this arrangement allows ship designers little flexibility in layout. In the superconducting system the engine that drives the generator can be located almost anywhere on the ship. And with fewer mechanical parts the drive takes up less total room, which would be especially important on board a submarine, where the extra space would be filled with weapons. In 1980 the navy launched the Jupiter II, the first vessel in history to be completely propelled by a superconductive drive.

Efficient superconducting motors and generators also have many important civilian applications. John Hulm headed one project at Westinghouse to develop a practical superconducting generator that could be used by the utilities. Ordinary generators are already very efficient; superconductivity would increase their efficiency by only about 1 percent, but even that small improvement is worthwhile. Over the lifetime of the machine, which is usually around thirty years, a superconducting generator would pay for itself many times over. But Westinghouse abandoned its superconducting generator program in 1983 because the amount of electrical power consumed in the United States was decreasing, and utilities were buying few generators of any kind. In Japan, on the other hand, which has no energy resources of its own, the development of superconducting generators is still being pursued vigorously.

Replacing conventional power lines with superconductors is another way to save energy. Again, the existing system is extremely efficient: Only about 5 percent of the energy generated at the power plant is lost on its trip to the consumer. But for conventional transmission lines this power loss is proportional to the length of the cable, which means that most electrical power is consumed within a few hundred miles of where it is generated. This is why nuclear power plants are often built near large population centers and not on more remote and safer sites. A prototype superconducting transmission line that used low-temperature superconductors was built at Brookhaven National Laboratory in Upton, New York, in the 1970s, but the cost of replacing existing transmission lines with liquid helium—cooled superconducting lines was too high, and the project was abandoned. But a transmission line that needed only liquid nitrogen for cooling, or did not need any cooling at all, could change the way electrical power is distributed and consumed.

The most promising energy saving application of superconductivity is known as superconducting magnetic energy storage (SMES). The rate of power consumption in the United States varies by as much as 50 percent in the course of a day; at night less energy is consumed than during the day. The utilities have to have sufficient generating capacity to handle the peak load, but when consumption drops this generating capacity goes to waste. In the case of fossil fuel–fired generators the plant sits idle while the utilities fret over all that capital investment that is not producing any return. Nuclear power plants *must* run at full output around the clock, and hydroelectric plants generate electricity as long as the water flows. Solar energy, of course, can be collected only during daylight hours.

The utilities have devised several schemes to bank some of the energy for use during times of peak consumption. One way that is actually used is to pump water uphill using the surplus energy that goes unconsumed during the night; when the energy is needed during the day the utilities let the water run back down the hill and turn a generator. This system, which is known as pumped hydropower, accounts for about 2.5 percent of the United States' energy-generating power. The main drawback is that it is inefficient; about a quarter of all the energy that goes in never comes out. In addition, pump hydropower plants must be located in the mountains, far from where the energy is consumed. Transmitting the stored power this extra distance may waste up to another 5 percent of the energy.

SMES is essentially nothing more than a giant superconducting magnet, that is, a fancy, looped superconducting wire. Once started current will flow forever around a superconducting loop. The energy required to start the current flowing is stored in the magnetic field, and can be tapped at any future time. The efficiency of SMES, according to Robert Loyd, an engineer who designs SMES facilities for Bechtel, is about 95 percent (losses come from converting A.C. current from the generating plant to the D.C. current that is stored in the coil, and vice versa, and from the energy the refrigeration system consumes). Bechtel has designed a 5000-megawatt-hour SMES facility using conventional metallic superconductors that can store or discharge energy at a rate of 1000 megawatts for five hours. It consists of a coil of superconducting wire over a half mile in diameter. The 1000 miles of superconducting wire from which this coil is wrapped will have to consist of a single, unbroken strand, because, once broken, superconducting wire cannot be repaired without an energy loss at the splice. Bechtel estimates that this facility would cost about $900 million, or about the same as a pumped hydropower plant of equal capacity.

The Pentagon is also interested in superconducting magnetic energy storage for use in Star Wars systems. For this application, energy stored over a long period of time could be released quickly, providing a deadly jolt of power that could be used to drive a ground-based laser or to launch missiles from a superconducting magnetic gun. Roger Boom of the University of Wisconsin, a proponent of such systems, told the *Wall Street Journal* that the energy a conventional generator takes fifteen minutes to store could be released in thirty-thousandths of a second. The Pentagon has recently awarded grants to two competing companies to design SMES systems. Boom suggests that energy stored by SMES systems built by utilities for their own use could be diverted to the Star Wars defense system in the event of a missile attack.

Having no resistance makes superconductors useful in a wide range of energy-saving applications. But in 1962 a brilliant young Cambridge graduate student named Brian Josephson predicted that superconductors should display another strange property as well. This property, now known as the Josephson effect, forms the basis for a wide range of electronic applications such as supersensitive magnetic sensors and tiny, superfast computers. Such devices have already been built from low-temperature superconductors. But, free of the intricate and expensive liquid-helium refrigerators they will find a far wider market and become as pervasive in future electronic applications as silicon is today.

Josephson was just beginning his graduate career when he predicted the effect that bears his name. He was doing experiments at the Mond Laboratory and taking a course in solid state physics from Philip Anderson, who was visiting Cambridge for a year. "This was a disconcerting experience for a lecturer, I can assure you," Anderson recalls, "because everything had to be right or he would come up and explain it to me after class."

One day, not long after Anderson had finished teaching his class about superconductivity, the twenty-two-year-old Josephson handed Anderson a paper he had just written. In it he analyzed a device of his own devising, which is now universally known as a Josephson junction: two superconductors separated by a thin insulating strip. Josephson showed that, according to the strange logic of quantum mechanics, superconducting pairs of electrons could leap over this strip, a balletic move that is forbidden by the laws of classical physics. Anderson could find no mistake in Josephson's reasoning. He was impressed. It was a neat idea, a macroscopic demonstration of a quantum mechanical effect. More than that, the electronic properties of Josephson junctions would soon prove to be useful, forming the basis of an entirely new discipline of superconducting electronics.

As soon as Anderson returned to Bell Labs he joined forces with an experimental physicist named John Rowell, who is now a vice-president of the Bell Communications Research Laboratory (Bellcore), and built the first working Josephson junctions. In 1973 Josephson was awarded a Nobel Prize. He was thirty-three, and with no new worlds left to conquer, at least as a physicist, he turned his attention inward and became a transcendental meditator, a disciple of the Maharishi Mahesh Yogi. Josephson tried to simulate the mind of the Maharishi on a computer. He also explored levitation—psychic, not superconducting—and has been depicted floating several inches above the floor on posters advertising transcendental meditation.

One of the simplest and most useful devices that can be made from Josephson junctions is called a superconducting quantum interference device, or SQUID. A SQUID is a superbly sensitive detector of magnetic fields, and has both civilian and military applications. And SQUIDs, which consist of just two Josephson junctions connected in parallel, are relatively easy to build. In fact, working SQUIDs using liquid-nitrogen-temperature superconductors have already been built by several groups in the United States and in Japan; they will probably be used in the first commercial applications of superconductivity.

Since the ocean is virtually transparent to magnetism, liquid-nitrogen-temperature SQUIDs will probably be used by the navy to detect submarines and mines. Geologists will use the new SQUIDs to map the Earth's magnetic field to locate subterranean mineral deposits.

Laboratory applications of SQUIDs also abound (in fact, low-temperature SQUIDs are even used by scientists who are looking for high-temperature superconductors). Medical researchers use SQUIDs to detect the minute magnetic fields generated by the tiny currents deep within the brain. The technique, which is known as magnetoencephalography (MEG), has many advantages over the two other leading methods of monitoring the brain. In electroencephalography, electrical currents generated deep within the brain are measured by electrodes on the scalp but the currents are distorted by their passage through skin and bone. Magnetic fields are only distorted by metal. The other leading method, called positron emission tomography (PET), creates its picture of the brain's activity by following radioactively tagged glucose as it is metabolized by the brain. The radioactive glucose is expensive and difficult to prepare and exposes the patient to potentially hazardous doses of radiation.

MEG is an entirely safe technique: The patient is not exposed to radioisotopes or to X rays. The apparatus does not even touch the patient. The few MEG machines that have been built using low-temperature SQUIDs have so far mostly been used for research. They consist of a small number of SQUIDs that must be moved around to map the magnetic field surrounding the patient's head. The more SQUIDs simultaneously monitoring the brain, the more detailed the scan. But low-temperature SQUIDs must be encased in bulky insulation. High-temperature SQUIDs would probably require far less insulation, allowing more of them to be clustered around the patient. When this becomes possible MEG will become an important clinical tool for diagnosing subtle brain abnormalities, such as epilepsy, dementia, stroke, and even schizophrenia or manic depression.

Josephson junctions can also be used as ultra-fast electronic switches from which it might be possible to build powerful, tiny computers. Computers, even those based on silicon, are already so fast that their designers must reckon with limitations due to the finite speed of light. In one nanosecond—one-billionth of a second, a long time in the supercomputer world—a beam of light travels only about three feet. So the faster a computer gets, the smaller it must be. The problem with this is that, ordinarily, the semiconductor switches that make up a computer generate heat. Not much heat, but when millions of them are squashed into a small volume, it adds up. Supercomputers like the speedy Cray need entire cooling plants to keep them from melting. Some conventional supercomputers already use liquid nitrogen for cooling.

Superconducting switches made from Josephson junctions dissipate hardly any heat at all. Moreover, they are fast; their switching times are measured in picoseconds—trillionths of a second. In the 1960s three large U.S. corporations—AT&T, IBM, and Sperry Univac (which later merged with Burroughs to form Unisys)—began research aimed at the construction of a superconducting computer.

The problems of making reliable microscopic Josephson junctions and wiring them together to form computer chips were formidable. Techniques of depositing thin films of superconductors in intricately patterned layers, similar to the techniques used to make silicon-integrated circuitry, had to be developed. Entire new circuit designs and computer architectures had to be invented. In 1979 Bell Labs dropped out; Sperry followed in 1983. IBM, which had the most ambitious program, was spending about $20 million a year in the early 1980s. The National Security Agency, which needs ever faster computers to break codes, among other things, contributed about $5 million a year to the IBM effort.

The computer logic chips the IBM scientists built

worked reasonably well, but the memory circuits proved to be more difficult to make than they had anticipated; they estimated that the memory problems would delay the program by at least two years. In the meantime, computers made from silicon and gallium arsenide were growing constantly faster. It was like trying to hit a moving target, the IBM management said. Someday, superconducting computers might be faster than conventional computers made from semiconductors, but that day was far off. So, in 1983, following the strict calculus of profit and loss, and despite pleas to continue from the National Security Agency, IBM terminated its superconducting computer program.

Saged Faris, who describes himself as "just an immigrant from the Sahara" (his childhood was spent in Libya), worked at IBM for eight years on developing superconducting computers. When IBM abandoned the project in 1983 Faris quit to seek venture capital funding to start a company of his own called Hypres, Inc.

Faris believes that superconductivity represents the "third electronics revolution." The first electronics revolution was initiated in 1904 by the invention of the "thermionic valve," or vacuum tube, by John Fleming (who, as has been noted, collaborated with James Dewar in measuring the resistance of materials at low temperature, the line of research that led to the discovery of superconductivity); the second electronics revolution was begun by the invention of the transistor by John Bardeen, William Shockley, and Walter Brattain (Bardeen, of course, is the "B" in BCS theory of superconductivity). The invention of the Josephson junction was the event that started the third electronics revolution.

Faris founded Hypres to spearhead that revolution. His

strategy is to start with a small application, and use the expertise developed along the way to bootstrap to ever more ambitious projects. In 1986 Hypres began marketing a piece of high-tech test equipment known as a sampling oscilloscope, a device that records fleeting events. At the heart of this device is a tiny glass slide on which scores of Josephson junctions are layered. The briefest event that can be recorded with a conventional electronic sampling oscilloscope is about 20 picoseconds; the Hypres oscilloscope is ten times faster, and the next models will be faster still.

This incremental, bootstrapping approach to commercialization is only possible—in the United States, at any rate—at a small, venture capital startup like Hypres. The rapid success of companies like Apple Computer or Compaq belies the fact that most venture capital is risked by people who are willing to wait seven to ten years for a possibly huge payoff. "We look at an area that most corporations steer away from," says George McKinney III, who is a partner in American Research and Development, a venture capital firm that has spawned such successes as the Digital Equipment Corporation. "Most corporations have a very short-term orientation, and a feeling that they can make up the long-term with investments or acquisitions. And I think they're correct. We look at a decade."

On March 19, 1987—the day after the "Woodstock of physics"—McKinney had breakfast with John Preston, a physicist with an MBA who is in charge of licensing companies to use inventions made at MIT. Most of the licenses issued by Preston are to companies that have fewer than twenty-five employees, small venture startups. Not surprisingly, most of these companies are started by MIT professors. Earlier in March, two MIT professors, John Vander Sande and Gregory Yurek, came to Preston with a method for making flexible "microcables" from the brittle ceramic superconductors. The idea was still in its embryonic stages, but the two scientists were confident that with a little time and money their method could be used to make cable for magnet and power applications. They filed some patents and then Preston reached for the phone. "Hey, McKinney, we've got a live one," he said.

By April 1, McKinney and the MIT professors had launched a company, American Superconductor, the first venture startup in America devoted to high-temperature superconductivity. By November McKinney and American Research and Development had raised $4.35 million, enough to hire seven employees and establish a 3000-square-foot development lab in Cambridge—the proverbial "garage" from which great technological empires spring. American Superconductor hopes to be producing cable suitable for demonstration tests somewhere between 1990 and 1992. "Only by 1995-plus," McKinney says, "will we see major production installations in the utility and magnet applications that justified the formation of American Superconductor."

In Japan, large corporations, backed by the Ministry of International Trade and Industry, are more willing to take on the kind of long-term research that in America only small venture firms like American Superconductor will risk. MITI's Electrotechnical Laboratory and Nippon Telephone & Telegraph both began Josephson junction research programs in the middle 1970s and those programs continue today. In 1980 MITI launched $100 million with the goal of building a superconducting computer capable of executing 10 trillion arithmetic operations a second— 10 gigaflops—by 1990. In this effort the Electrotechnical Laboratory's ongoing effort was joined by Hitachi, NEC, and Fujitsu. "In 1983 IBM halted their research on the Josephson junction computer," Katsuzo Aihara, a senior researcher at Hitachi, recalled. "Our company decided to continue. Of course, there were many problems, but we thought that in ten or twenty years it would be practical, so we continued."

The Japanese, benefiting from U.S. Josephson junction R&D, have made impressive progress toward MITI's goal. Fujitsu recently unveiled the world's first superconducting microprocessor, which is ten times faster than an ordinary microprocessor and dissipates four hundred times less heat. Saged Faris estimates that the Japanese are at least four years ahead of the United States. And the expertise the Japanese have developed working on low-temperature superconductivity should, at least in part, translate to the high-temperature ceramics.

Japan's penchant for long-term commitment and planning is also illustrated by the story of their devotion to another application of superconductivity. In this case, again, they were inspired by the United States, and again they persisted long after the United States had lost interest.

In the 1960s a visionary American scientist at MIT named Henry Kolm had an idea that he felt sure would revolutionize transportation: a train that would race between cities at 200 or 300 miles an hour. Ordinarily the speed of a train is limited by the vibration and friction of its wheels against the track. Kolm said, Get rid of the wheels—let the train be supported on a powerful magnetic field. Magnets would also propel the train, dragging it down the track in an arrangement that Kolm called a linear motor. Kolm dubbed his invention the "magnaplane"; now his idea is more generally called "maglev," short for magnetic levitation. "A maglev transportation system will alleviate one of the most serious crises of our time," Kolm said, "traffic congestion. The magnaplane can carry twenty thousand people at two hundred miles per hour." Maglev projects were undertaken in the United States, West Germany, and Japan. In 1975 the United States abandoned its maglev project.

The Japanese National Railroad had begun its maglev project in the early 1960s, and has steadily worked on it ever since. Its goal was a train that could allow workers to commute from Tokyo to Osaka, about 300 miles, in one hour, allowing a radical redistribution of Japan's already large, and growing, population. It was social engineering on a scale only imaginable in Japan, and the Japanese have pursued it for over twenty-five years.

In 1969 the Japanese engineers decided to use superconducting magnets to levitate the train, because their modest power consumption obviates the need to supply energy from the ground. Onboard power is only needed to keep the magnets cold; liquid-nitrogen-cooled magnets would be a lot easier to refrigerate. Conventional, resistive magnets built into the track speed the train on its way. To design the superconducting magnets and refrigeration systems the Japanese enlisted the help of Toshiba, Mitsubishi Electric, and Sumitomo Heavy Industries—and as a result of their participation these companies gained valuable experience in the design and construction of low-temperature superconducting magnets, some of which should be directly applicable to the new materials.

In 1987, on the southern island of Mimitsu in the Miyazaki Prefecture, Kysushu, the Japanese demonstrated a prototype train called the MLU-002, capable of carrying forty-four passengers and traveling at over 300 miles per hour. At low speeds the train is supported on rubber wheels akin to landing gear. But as the train rapidly accelerates to its cruising speed the wheels are withdrawn into the body and the train climbs to its cruising altitude of about four inches above the track; the transition from rolling to flying is virtually imperceptible, but the ride is bumpier than the Shinkansen, the more conventional Japanese bullet train that travels 125 miles per hour.

Hiroshi Takeda, chief researcher on the maglev project for Japan Railway, estimates that Japan has so far spent ¥43 billion ($330 million) on the project. The next step will be to build a 30-mile-long test track at the cost of another ¥150 billion ($1.2 billion), but Takeda says that this step might be skipped. The track, excluding stations, will cost about ¥2.5 billion per kilometer ($32 million per mile). Takeda claims that there are no further technical problems to solve—the rest is politics. Takeda optimistically expects to make the one-hour run from Tokyo to Osaka by the end of the century.

The Japanese are hoping to export their superconducting maglev technology around the world. Their only competition comes from a West German consortium called Transrapid International. The German train does not use superconductors at all, so the onboard magnets require a steady supply of electrical current. As a result it can only raise itself about an inch above the track, a quarter of the height of the Japanese model, which makes it more vulnerable to earthquakes—a consideration more important in Japan than in Germany.

With their long-standing and continuous interest in superconductivity the Japanese are in a good position to rapidly commercialize the new superconductors. After Shoji Tanaka's group at Tokyo University confirmed Bednorz and Müller's initial discovery—first announced in the November 28, 1986, issue of *Asahi Shimbun*—hundreds of articles about superconductivity have appeared in Japanese magazines and newspapers. Best-selling books about the "superconductivity revolution" appeared almost immediately, as did comic books in which young men are rejected by attractive women for not knowing about superconductivity. More important were the hundreds of scientific papers and patents that immediately began flowing from Japan's laboratories. By April it was rumored that Sumitomo Electric alone had taken out 700 patents; even allowing for the fact that Japanese patents are narrower than those issued in the United States so that seven Japanese patents are roughly equal to one U.S. patent, the number was impressive and even a bit disconcerting. By the beginning of 1988 the Japanese had already applied for approximately 2000 international patents concerning high-temperature superconductivity; the filing fee is $10,000 per patent, so these patents alone represent a $20 million investment.

The OTA assessment of the status of the race toward commercialization is somewhat pessimistic (they dismiss the significant but smaller efforts in Europe, the Soviet Union, and China in a footnote): "American companies may already have begun to fall behind. Japanese firms have been much more aggressive in studying possible applications of high temperature superconductivity, and have more people at work, many of them application-oriented engineers and business planners charged with thinking of ways to get high-temperature superconductivity into the marketplace." (In the final days of the Reagan administration a White House panel of "wise men" reached a similar conclusion. "I think people are seeing there is a gap," Ralph Gomory, senior vice president for science and technology of IBM and chairman of the panel, told the *New York Times* in January, 1989. To narrow this gap the committee recommended increasing the funding of basic university research by several million dollars a year and establishing four to six Japanese-style consortia.)

The Japanese are especially good at materials processing and manufacture, areas in which the United States is particularly weak. In particular, the Japanese have considerable experience in processing advanced ceramic materials, which should translate to the processing of ceramic superconductors.

"Most Japanese managers believe high temperature superconductivity to be closer to the marketplace than do their American counterparts," the OTA observed. "Seeking growth and diversification, they have assigned more people to high temperature superconductivity than U.S. firms, and may also be spending more money. The Japanese have committed funds, not only to research but to evaluating prospective applications."

Even a company like Matsushita—known as Pioneer in the United States—which only makes consumer products, has taken an interest in superconductivity. Since consumers are unlikely to buy products that need to be filled with liquid nitrogen, Matsushita is concentrating most of its effort on finding room-temperature superconductors. Having begun this line of research, how long is Matsushita willing to wait for a payoff? The answer might make an American manager cringe: "Once Matsushita starts a research project we don't give up until we have success," a company spokesman said. "Now we are going into the battlefield."

In 1988 the size of the Japanese war chest—the amount of money the government spent on superconductivity— came to about $57 million, an increase of 300 percent in one year. This figure does not include salaries, so it is hard to compare to the U.S. budget, which in 1988 came to about $147.5 million, a gain of 75 percent. In neither country will money be the problem. More important will be the less quantifiable factors, such as commitment, experience, and education, three areas in which the Japanese are particularly strong.

A novel aspect of the Japanese government's superconductivity program is that, at least in theory, it stresses international cooperation. A key part of the MITI program was the establishment of an International Superconducting Technology Engineering Center (ISTEC); foreign companies have been invited to join. The membership fee comes to around $750,000, with yearly dues another $100,000. In addition, member corporations are required to send two scientists to Japan to work year-round in ISTEC's Tokyo research center. For a smaller investment—a $15,000 initiation fee, plus dues of $15,000 per year— companies can have limited access to ISTEC research. So far no foreign company has become a full member in ISTEC and only five companies—four American and one British—have become limited members. The reason most companies give for not joining is the expense. Moreover, they claim not to see what the advantages are.

Shoji Tanaka, who has left Tokyo University to head ISTEC, thinks that the Americans are missing the point. "Superconductivity is like a precious stone in the earth," Tanaka says. "We must develop it carefully. It must not be dominated by one person, one company, or one country." He is afraid that the snubbing of Japanese overtures of cooperation by the United States will lead to a bitter technological war, increasing tensions between the two countries.

Should America fear the Japanese? "Oh, absolutely," says Kent Bowen, an MIT professor with close ties to Japan. "Fear, not in the sense that they're evil. They're just good. It points up our weaknesses." Bowen sees the competition over superconductors as an opportunity "to learn how we as Americans, with our institutions and our good points and our bad points, how we can learn to do it right. The Japanese do their thing very well; can we learn to do our thing that well?"

One of the things that the Japanese excel at is making incremental improvements to existing technology. A few years back people began to investigate replacing the silicon of computer chips with gallium arsenide. Gallium arsenide circuitry is faster than silicon, but the necessary techniques of growing single crystals of gallium arsenide had not been developed. Americans had invented a method called abridgment, but it was a mess; they began to look for a better, more elegant method.

Meanwhile, Sumitomo Electric, a manufacturer of electrical cable—a company not noted for its high-tech brilliance—decided to see if it could make the messy, American method of abridgment work. They set up a pilot production line, Bowen says, "and just started making it.

And making it, and making, and making it. And bigger and bigger, and they'd have to waste a lot of time and material on that early stuff, but they began to learn to make it. And we were waiting—I don't know—for miracles, for something. The home run. They're the bunts and singles guys over there, and we're the home run hitters."

There are lessons to be learned in both America and Japan from carefully observing the superconductivity race. Experience has shown that, on the average, it takes from ten to twenty years for a laboratory discovery to become a commercial product. The first products containing high-temperature superconductors could be on the market, by some estimates, in two or three years. This will happen only if government and industry and society discover new ways of managing innovation. This discovery, Bowen believes, would be even more important than the discovery of high-temperature superconductors. Bowen and colleagues at the Harvard Business School are watching the superconductivity race with a disinterested fascination. "Let's assume that we sat back and analyzed [the new process of innovation] and we were able to write down the paradigms for it, and we were able to bottle it and give it to people every time there's a new scientific discovery . . . It totally changes the world in which we live!"

CHAPTER 9

ANNUS MIRABILIS

Though few would admit it, October is a nervous month for many of the world's top scientists. October is when the call from the Nobel Foundation in Stockholm either comes or it doesn't. Good form dictates, as journalist Tom Wolfe has written, that "anyone who entertains the notion of a Nobel Prize is advised to act as if he is oblivious of its very existence." In apparent compliance with this advice, Alex Müller was attending a conference in Italy when the call came informing him that he and Georg Bednorz had been awarded the 1987 Nobel Prize in physics for their discovery of the new class of high-temperature superconductors. IBM promptly sent a corporate jet to return Müller to Zürich to celebrate their triumph.

Bednorz and Müller's Nobel Prize was the second to be earned in as many years by physicists at IBM's Zürich Research Laboratory, which employs only about 200 of IBM's 3000 research scientists. The previous October, Gerd Binnig and Heinrich Rohrer learned that they would share in the 1986 Nobel Prize for their invention of the scanning tunneling microscope with which it is possible to create images of individual atoms. So in 1987 IBM became only the second corporate laboratory in history to win back-to-back Nobel Prizes (the other was Bell Labs; three nonprofit or university labs have also turned this trick).

Few questioned the swiftness with which the Nobel Committee recognized Bednorz and Müller's contribution. Their work had been duplicated scores of times by researchers around the world and led the way to a stunning series of discoveries that would undoubtedly benefit mankind; the only real question was when and to what extent. But the Nobel Prize can be shared by as many as three persons, and some felt that Paul Chu should have been included.

The Nobel Committee conducts its deliberations with the kind of secrecy that the Vatican employs in choosing a pope, so the reasoning behind their excluding Paul Chu will probably never be known. Nevertheless, those who watch for the telltale smoke have suggested a number of plausible explanations. They point out that Chu's discovery was made in 1987, and the rules stipulate that the Nobel Prize cannot be awarded in the same calendar year in which the work was done. Chu could not have shared in a 1987 Nobel Prize. Of course, the Nobel Committee could have delayed giving the prize for a year, but, given the astonishing interest which the public had shown for superconductivity, it would have seemed odd if the Nobel Prize were given for some relatively obscure achievement.

Even if Chu had been allowed to share in the prize, some argue that his discovery did not deserve it. The discovery of liquid-nitrogen-temperature superconductors was important, but after Bednorz and Müller's discovery had become known it was also probably inevitable. Many groups were performing the same kind of substitutions as Chu's group and sooner or later one of them would have hit upon substituting yttrium for lanthanum. (In fact, as it came out only much later, the idea of using yttrium was conceived and tried at the University of Alabama by Chu's collaborator and former student, Maw-Kuen Wu. For more than a year Wu's contribution had been virtually ignored by the press and the scientific community, a misunderstanding that Chu apparently made little effort to correct. The Nobel Committee's wise decision to recognize only the irrefutable contribution of Bednorz and Müller in 1987 assured that justice was done: It would have been unfair to have excluded Wu from sharing in the prize if Chu had gotten it; since the Nobel Prize can be shared by only three scientists neither Chu nor Wu could share in it.)

While a Nobel Prize has eluded Chu, he has been honored in the United States, where he has become a "superstar scientist." Edward Teller, father of the hydrogen bomb, traveled to Houston just to meet Chu. Teller scorned the Nobel Committee's decision, and pleased his host, when he contrasted "the minor Zürich oxide and the major Texas oxide." (But then, to Teller the atom bomb was just a minor bomb.) Chu became one of the panel of "Wise Men" chosen to guide President Reagan's Superconductivity Initiative. He won the National Medal of Science. The University of California, Berkeley, tried to lure him away from the University of Houston. The attempt failed when some proud Texans endowed a chair for Chu that paid $150,000 a year, and set him up in his own new Texas Center for Superconductivity. Don Henderson, a Republican state senator from Houston, said that the state's initial contribution of $1.5 million for the center was "literally the only money appropriated during the legislature's regular session, which failed to produce a state budget."

Wu, who may have received less public attention than

he deserved, nevertheless was rewarded for his part in the discovery. At the time of his collaboration with Chu he was an assistant professor without tenure at the University of Alabama; now he is a tenured full professor at Columbia University. There are rewards in science other than the Nobel Prize.

In the year that followed the discovery of 1-2-3, the 90-K superconductor, scientists all over the world baked up thousands of compounds in the hope of finding a new material with a higher critical temperature, but met with no success. This was discouraging not only to those devoted to the commercialization of superconductivity, but to the theorists who were attempting to uncover the mechanism of high-temperature superconductivity. With only a few known high-temperature superconductors it was difficult to make the kinds of generalizations that lead to a good theory—just as it is difficult to break a code given only a few short messages. By the end of 1987 some scientists, new to the field of superconductivity and spoiled by the early pace of discovery, began to grow pessimistic. It seemed possible that after an initial flurry of discoveries superconductivity would once again stall as it had in the 1970s. "I would guess somebody would have [made a new material] if it was going to be done," said Eric Forsyth of the Brookhaven National Laboratory. "It seems as though they have evaluated just about every possible permutation and combination of these new ceramic superconductors." Chu took a more philosophical view. "If breakthroughs could occur every day," he said, "our feeble hearts could probably not handle the stress."

The apparent slowing of the pace of discovery was an

illusion; the stakes of the game had suddenly become enormous, and researchers were writing patent applications long before embarking on scientific papers. In May 1987 a couple of researchers at the University of Arkansas, Allen Hermann, the co-inventor of the lithium-iodine battery that is used in coronary pacemakers, and his research associate Zhengzhi Sheng, joined the international hunt for new superconductors. In October, after having been enlightened by a long string of failures, the two researchers thought of substituting thallium for yttrium. Their first batches of thallium-barium-copper oxide became superconducting at a temperature of about 82 K. It took three more months for the University of Arkansas lawyers to go through the lengthy patent process, and during this period Hermann and Sheng were not allowed to say anything. Finally, on January 22, 1988, they held a press conference at the University of Arkansas to announce their discovery.

The news conference was not the stunning success it might have been, because on that same day a group at Japan's National Research Institute of Metals, led by Hiroshi Maeda, announced that they had discovered a new group of superconductors made from bismuth, barium, calcium, copper, and oxygen. The Japanese superconductor occurred in at least two different phases with critical temperatures of 75 and 105 K. Paul Chu was stunned by the Japanese announcement, because his group had made a very similar discovery that they had been keeping secret for some time. "They were a few days ahead of us making the announcement," Chu said. It didn't really matter. Hoechst AG, a West German pharmaceutical manufacturer, had filed for a patent on the bismuth material the previous November. The patent had been filed in West

Germany, the United States, and Japan, but Hoechst had decided to keep quiet about it for (unspecified) commercial reasons.

Within two weeks of their unheralded January announcement Hermann and Sheng had seasoned their brew of thallium, barium, copper, and oxygen with calcium, and the critical temperature soared to a record of nearly 125 K. They held another press conference on their record-breaking discovery in early February, but few reporters paid attention. Most, in fact, were in Boston, attending the annual meeting of the American Association for the Advancement of Science. The talk at that meeting was mostly focused on the bismuth superconductors. At a press conference held by Philip Anderson and other important and visible superconductivity researchers the thallium compounds were not even mentioned. Paul Raeburn, an Associated Press reporter, attended this conference and promptly filed a story about the bismuth compound. The next day his editor in New York called and said, "We have a short story here from an Arkansas newspaper about a superconducting material that is supposed to be pretty good." Raeburn told his editor, "It can't be anything. I just heard four or five of the top people talk about everything new in superconductivity. They didn't say anything about it. Forget it. Kill it." Having in the past year written many stories about amazing new materials that later turned out to be mistakes, Raeburn and other reporters had become cautious. As a result the thallium compound, which has the highest known transition temperature, as well as other properties that make it a leading candidate for speedy commercialization, was hardly mentioned at all in the popular press.

A few days after their second news conference Hermann and Sheng unveiled their discovery to an audience of their peers. They divulged details of their formula at a 5 P.M. poster session—a kind of professional science fair—during the World Conference on Superconductivity on February 21, 1988. By 9:30 the following morning Du Pont had announced that they had confirmed the Arkansas result; Paul Chu's confirming nod came that afternoon; the Japanese took until the next day.

The headlong rush to explore the new Arkansas materials was slowed somewhat by thallium's infamy as one of the most toxic of all the elements. When one researcher was asked whether thallium causes cancer he said, "Nobody knows because it kills you first." Chu said that some of his students did not want to work with the stuff; one even threatened him with a lawsuit. Georg Bednorz, who could well afford to be fastidious, said that he would not work on the thallium-based materials. "I'm not going to kill myself for five degrees," he put it. Arthur Sleight of Du Pont was more philosophical. "Nature has mechanisms to limit populations, and we do have a highly overpopulated field," he said. "Perhaps thallium has come along to solve that problem." Nevertheless, with proper precautions thallium can be handled safely in the laboratory, and once baked into a superconductor it becomes relatively innocuous.

The new bismuth and thallium superconductors are both proving to be easier to form into wires and films with desirable technological properties. But the very existence of these two new systems is perhaps even more important than their potential for commercialization; it proves that high-temperature superconductivity is a general phenomenon not restricted to one very narrow class of materials.

With only one system it was hard to tell which features of the crystal structure were important to their superconductivity and which were superfluous. The crystal structures of all three systems are similar in that they all contain corrugated planes of copper and oxygen atoms sandwiched between layers of the other elements. The implication is clear: superconductivity takes place in these copper-oxygen planes. Furthermore, it seems that materials containing more adjacent copper-oxygen planes have higher transition temperatures. With this clue researchers have been trying to find ways of synthesizing new materials with more planes in the hopes of pushing up the critical temperature. Room temperature, of course, is still the ultimate goal, but smaller increases would also be technologically important. For example, a temperature increase of only about 20 K over the current record critical temperature of 125 K would bring superconductivity above the boiling point of liquid freon, the coolant used in conventional refrigerators.

In one brief, miraculous year more was learned about these brittle, dusky superconductors than virtually any other material in the history of science. An entirely new field, with its own journals and jealousies, heroes and legends, was formed with the rapidity of a volcanic island. How high-temperature superconductors work remains a mystery, but that they work, and that they will someday form the basis of whole new technologies, seems certain —if for no other reason than by dint of the determination and ingenuity of thousands of men and women working in laboratories in every part of the world. "In the long run," Thoreau wrote, "men only hit what they aim at."

The fact of high-temperature superconductivity must be factored into every projected view of the future. Plentiful energy, flying trains, faster computers now all seem possible—and in technology what is possible is inevitable. But the most important consequences of superconductors may be found in the realm of the imagination.

When Einstein was a boy of five and briefly confined to bed with an illness, he received what his biographer Ronald W. Clark called "the first genuine shock to his intellectual system." To amuse his bedridden son, Einstein's father brought him a magnetic compass. The child was mesmerized by the mystery of the quivering needle that, no matter where he hid the compass or how the case was turned, always pointed north. That mystery drew Einstein inevitably toward the studies of space and time that revolutionized man's conception of the universe. What revolutions will some future Einstein conceive, having once seen a silvery magnet hanging placidly in the space above a dark, superconducting wafer?

SOURCE NOTES

1 SURMISES, JEALOUSIES, CONJECTURES
This chapter is based primarily on interviews with Georg
Bednorz, Ted Geballe, Koichi Kitazawa, Alex Müller, and
Shoji Tanaka.

2 THE GENTLEMAN OF ABSOLUTE ZERO

Armstrong, Henry David. 1924. "James Dewar; a Friday Eve-
    ning Lecture to Members of the Royal Institution on
    January 18, 1924."
Crowther, J. G. 1968. *Scientific Types.* Dufour Editions, Inc.
Dewar, James. *Encyclopaedia Britannica,* 11th edition,
    1910, "Liquefaction of Gases."
Dictionary of Scientific Biography. 1973. Charles Coulston
    Gillespie, editor in chief. New York: Charles Scribner's
    Sons.
Gough, William C. 1967. "Birth of a New Technology: Super-
    conductivity." Research paper presented to the Science
    and Public Policy Seminar, Harvard University (unpub-
    lished).
"Heike Kamerlingh Onnes 1853–1926." *Proceedings of the
    Royal Society.* A 113 (January 1927).
Kamerlingh Onnes, H. 1908. Communication of the Physical
    Laboratory, University of Leiden (no. 108).

Kamerlingh Onnes, H. 1911. Communications of the Physical Laboratory, University of Leiden (no. 120b and no. 1246).

Kamerlingh Onnes, Heike. 1913. "Report on Researches Made in the Leiden Cryogenic Laboratory between the 2nd and 3rd International Conference."

Mendelssohn, Kurt. 1977. *The Quest for Absolute Zero.* New York: Ticknor and Fields.

3 SCHMUTZ PHYSICS

The material in this chapter derives largely from interviews with John Hulm, Ted Geballe, Brian Maple, Gloria Lubkin, and others who knew Matthias. In addition the following references were used:

Clogston, Albert M., Theodore H. Geballe, and John Hulm. 1981. "Bernd T. Matthias," *Physics Today* 34 (January), p. 84.

Mathews, W. N., W. D. Gregory, and E. A. Edelsack. 1973. *The Science and Technology of Superconductivity.* New York: Plenum Press.

Matthias, Bernd T., John K. Hulm, and Eugene J. Kunzler. 1981. "The Road to Superconducting Materials." *Physics Today* 34 (January), p. 34.

4 THE OUTSIDERS

This chapter is based largely on interviews with Alex Müller and Georg Bednorz, with reference to their published papers. Although they are not mentioned there by name, the first announcement in the United States of Bednorz and Müller's discovery appeared in the *New York Times,* December 31, 1986, Sec. 1, p. 1. Other accounts in the *Wall Street Journal* and *New York Times* were also important.

5 SERENDIPITY

Despite my repeated efforts, Paul Chu was the only scientist who would not grant me an interview for this book. I was able to hear him speak on numerous occasions and questioned him at press conferences, but I was forced to rely on interviews with others who knew Chu or had worked with him. In addition to those interviews, I relied on articles that appeared in the *New York Times* and the *Houston Post.* The following sources were particularly valuable:

Berryhill, Michael. 1988. "A Conversation with Paul Chu," *Superconductor World Report,* 2 (no. 2, March), p. 8.

Gleick, James. 1987. "In the Trenches of Science," *New York Times Magazine,* August 16, p. 28.

"NOVA: Superconductor." 1987. Produced by Linda Garmen, WGBH, Boston.

Poole, Robert. 1988. "Superconductor Credits Bypass Alabama," *Science,* 241 (August 5), pp. 655–58.

Superconductor Week. 1988. "An Interview with Robert Hazen," 2 (no. 15, April 18), pp. 4–5.

6 GREAT EXPECTATIONS
In addition to the interviews cited in the text, additional material was gathered from accounts of the March meeting that appeared in the *New York Times, Physics Today, Time,* and *Newsweek.*

Braun, Ernest, and Stuart MacDonald. 1978. *Revolution in Miniature; The History and Impact of Semiconductor Electronics.* Cambridge, England: Cambridge University Press.

7 THE RIDDLE
Bernstein, Jeremy. 1984. *Three Degrees above Zero.* New York: Charles Scribner's Sons.

Dictionary of Scientific Biography. 1973. Charles Coulston Gillespie, editor in chief. New York: Charles Scribner's Sons.

Kittel, Charles. 1968. *Introduction to Solid State Physics,* 3rd edition. New York: John Wiley and Sons.

8 SUPER-TECH
CSAC Washington Update, Council on Superconductivity for American Competitiveness newsletter, all issues.

Dickson, David. 1984. *The New Politics of Science.* Chicago: University of Chicago Press.

Federal Conference on Commercial Applications of Superconductivity, July 28–29, 1987. Transcripts prepared by *Superconductor Week* staff.

Focus on Japan, translations of articles that appeared in *Nihon Kogyo Shimbun* and *Nikkei Sangyo Shimbun,* published by the Council on Superconductivity for American Competitiveness.

Lardner, James. 1987. *Fast Forward: Hollywood, The Japanese and the VCR Wars.* New York: W. W. Norton & Company.

Schwartz, Brian B., and Simon Foner, eds. 1976. *Superconductor Applications: SQUID and Machines.* Cambridge, Mass.: Bitter Laboratory, M.I.T.

Sullivan, D. B., ed. 1978. *The Role of Superconductivity in the Space Program.* Prepared for NASA by the National Bureau of Standards (May) NBSIR 78–885.

Superconductor Week, numerous articles in vols. 1 and 2 (1987 and 1988).

U.S. Congress, Office of Technology Assessment. 1988. *Commercializing High-Temperature Superconductivity.* Washington, D.C.: U.S. Government Printing Office (June) OTA-ITE-388.

9 ANNUS MIRABILIS

This chapter is based on interviews. See also:

Clark, Ronald W. 1971. *Einstein: The Life and Times.* New York and Cleveland: The World Publishing Company.

INDEX